JAVA PROGRAMMING

Editors

M.PRAVEENA, MCA, M.Phil.,

B.NANDHINI, M.Sc., M.Phil.,

SHANLAX PUBLICATIONS

JAVA PROGRAMMING

First Edition: 2018

ISBN: 978-93-87871-31-1

Price: Rs.440/-

Publisher

SHANLAX PUBLICATIONS

66, T.P.K. Main Road
Vasantha Nagar
Madurai – 625003
Tamil Nadu, India

Ph: 0452-4208765,
Mobile: 7639303383
email:publisher@shanlaxpublications.com
web: www.shanlaxpublications.com

ACKNOWLEDGEMENT

This book is an introduction to Java programming for beginners. It is tailored for students preparing for the Computer Science, but it is for anyone who wants to learn Java. This is an easy-to-follow textbook that guides the beginning programmer step-by-step through the process of learning Java. This book helps you learn the language basics, AWT, Networking and some chapters on Servlet, JSP, plus covering some analysis. The main obstacle to learning object-oriented programming is the volume of interdependent detail that needs to be learned before even the simplest program can be created. This text eliminates extraneous details early on and stresses object concepts that will provide a basis for students to become expert programmers. Classes, objects, and working programs are introduced at the outset, and programming is presented as extended problem solving, making it easier to understand. Advanced Java Programming is the perfect text for anyone new to Java who wants a comprehensive, easy-to-comprehend reference.

M. Praveena is an Assistant Professor, Department of Computer Science at Dr. SNS Rajalakshmi College of Arts and Science, Coimbatore, India. She received the B.Sc (Che/Phy)., from Avinashilingam Deemed University in the year 2003, Coimbatore and MCA from Dr. SNS Rajalakshmi College of Arts and Science under Bharathiar University in the year 2006. She received the M.Phil. Degree from Periyar University, Salem in 2007. Her area of interest is Data Mining.

B. Nandhini is a Research Scholar, Department of Computer Science at Dr. SNS Rajalakshmi College of Arts and Science, Coimbatore, India. She received the B.Sc (CS)., from PSGR Krishnammal College of Arts and Science, Bharathiar University in the year 2014, Coimbatore and M.Sc (CS) from PSGR Krishnammal College of Arts and Science under Bharathiar University in the year 2016. She received the M.Phil., degree from Bharathiar University, Coimbatore in 2018. Her area of interest is Advanced Networking.

This book is an introduction to Java programming for beginners. It is tailored for students preparing for the Computer Science, but it is for anyone who wants to learn Java. This is an easy-to-follow textbook that guides the beginning programmer step-by-step through the process of learning Java. This book helps you learn the language basics, AWT, Networking and some chapters on Servlet, JSP, plus covering some analysis. The main obstacle to learning object-oriented programming is the volume of interdependent detail that needs to be learned before even the simplest program can be created. This text eliminates extraneous details early on and stresses object concepts that will provide a basis for students to become expert programmers. Classes, objects, and working programs are introduced at the outset, and programming is presented as extended problem solving, making it easier to understand. Advanced

Java Programming is the perfect text for anyone new to Java who wants a comprehensive, easy-to-comprehend reference.

I am very happy to present the book before the esteemed readers. The main aim of this book is to provide easy understanding of the concepts for the beginners. The topics covered in this book have been chosen keeping in view the fundamentals ideas required for the students of computer science. Examples have been given at appropriate places. It is my earnest hope that the student and teacher will find the book more useful. Suggestions for improvement of this book are warmly welcomed.

I would like to thank the publishers .... for their untiring efforts and whole-hearted co-operation in bringing out this book.

M.PRAVEENA, MCA, M.Phil.,

B.NANDHINI, M.Sc., M.Phil.,

CONTENT

CHAPTER-1

JAVA – OVERVIEW

1.1 What is Java?

- Java is an **object-oriented programming** language developed by **Sun Microsystems** and released in **1995.**
- Java was originally developed by **James Gosling at Sun Microsystems** (which has since merge into Oracle Corporation).
- Java programs are platform independent which means they can be run on any operating system with any type of processor as long as the Java interpreter is available on that system.
- Java code that runs on one platform do not need to be recompiled to run on another platform, it's called write once, run anywhere (WORA).
- Java Virtual Machine **(JVM)** executes Java code, but is written in platform specific languages such as C/C++/ASM etc. JVM is not written in Java and hence cannot be platform independent and Java interpreter is actually a part of JVM.

1.2 Where is Java being used?

Earlier java was only used to design and program small computing devices but later adopted as one of the platform independent programming language and now according to Sun, 3 billion devices run java.

- Java is one of the most important programming languages in today's IT industries.
- **JSP –** Java is used to create web applications like PHP and ASP, JSP (Java Server Pages) used with normal HTML tags, which helps to create dynamic web pages.
- **Applets –** This is another type of Java program that used within a web page to add many new features to a web browser.
- **J2EE –** The software Java 2 Enterprise Edition are used by various companies to transfer data based on XML structured documents between one another.
- **JavaBeans –** This is something like Visual Basic, a reusable software component that can be easily assembled to create some new and advanced application.
- **Mobile –** Besides the above technology, Java is also used in mobile devices, much kind of games and services built-in Java. Today, all leading mobile service provider like Nokia, Siemens, Vodafone are using Java technology.

1.3 Types of Java Applications

- **Web Application –** Java is used to create server-side web applications. Currently, servlet, jsp, struts, jsf etc. technologies are used.
- **Standalone Application –** It is also known as desktop application or window-based application. An application that we need to install on every machine or server such as

media player, antivirus etc. AWT and Swing are used in java for creating standalone applications.

- **Enterprise Application –** An application that is distributed in nature, such as banking applications etc. It has the advantage of high level security, load balancing and clustering. In java, EJB is used for creating enterprise applications.
- **Mobile Application –** Java is used to create application softwares for mobile devices. Currently Java ME is used for creating applications for small devices, and also Java is programming language for Google Android application development.

1.4 Facts about Java

- **Object Oriented –** In java everything is an Object. Java can be easily expanded since it is based on the Object model.
- **Platform independent –** C and C++ are platform dependency languages hence the application programs written in one Operating system cannot run in any other Operating system, but in platform independence language like Java application programs written in one Operating system can able to run on any Operating system.
- **Simple –** Java is designed to be easy to learn. If you understand the basic concept of OOP java would be easy to master.
- **Secure –** With Java's secure feature it enables to develop virus-free, tamper-free systems. Authentication techniques are based on public-key encryption.
- **Architectural-neutral –** Java compiler generates an architecture-neutral object file format which makes the compiled code to be executable on many processors, with the presence Java runtime system.
- **Portable –** being architectural neutral and having no implementation dependent aspects of the specification makes Java portable. Compiler and Java is written in ANSI C with a clean portability boundary which is a POSIX subset.
- **Robust –** Java makes an effort to eliminate error prone situations by emphasizing mainly on compile time error checking and runtime checking.
- **Multi-threaded –** With Java's multi-threaded feature it is possible to write programs that can do many tasks simultaneously. This design feature allows developers to construct smoothly running interactive applications.
- **Interpreted –** Java byte code is translated on the fly to native machine instructions and is not stored anywhere. The development process is more rapid and analytical since the linking is an incremental and light weight process.
- **High Performance –** With the use of Just-In-Time compilers Java enables high performance.
- **Distributed –** Java is designed for the distributed environment of the internet.
- **Dynamic –** Java is considered to be more dynamic than C or C++ since it is designed to adapt to an evolving environment. Java programs can carry an extensive amount of run-time information that can be used to verify and resolve accesses to objects on run-time.

1.5 Different Editions of Java Technology

- **Java SE –** Java SE or Java Standard Edition provides tools and API's that you can use to create server applications, desktop applications, and even applets. These programs developed using Java SE can be run on almost every popular operating system, including Linux, Macintosh, Solaris, and Windows.
- **JEE –** Based on the foundation framework of the standard edition, Java Enterprise Edition helps in web application service, component model and enterprise class service oriented architecture (SOA).
- **JME –** Java Micro Edition or JME for short is an accumulation of Java APIs that are used for the development of software for devices like mobile phones, PDAs, TV set-top boxes, game programming. The platform of micro edition generally consists of an easy user interface, a robust security model and a wide variety of built-in networks for running Java based application.

CHAPTER-2

HISTORY OF JAVA TECHNOLOGY

2.1 History of Java

- James Gosling, Patrick Naughton, Chris Warth, Mike Sheridan and Ed Frank initiated the Java language project in June 1991.
- The idea was to develop a language which was platform-independent and which could create embedded software for consumer electronic devices.
- The language took 18 months to develop and had an initial name as Oak which was renamed to Java in 1995, due to copyright issues. Originally developed by James Gosling at Sun Microsystems (which has since merge into Oracle Corporation) and released in 1995. JDK 1.0 released in (January 23, 1996).
- Java SE 8.0 is a current stable release of Java, and many other previous Java versions are also available.

2.2 Java Version Release History

Version	Released on	Description
JDK1.0	23 Jan 1996	The codename is Oak. Initial release of Jdk1.0
JDK1.1	19 Feb 1997	The addition of AWT, JDBC, RMI, Javabeans etc
J2SE 1.2	8 Dec 1998	J2SE(Java2 Standard Edition), Codename Playground, Major additions are- Swing graphical API was integrated, strictfp Keyword, reflection, Java IDL, collection and JIT Compiler in Sun's JVM for Java plug-in was equipped.
J2SE 1.3	8 May 2000	Codename Kestrel. The major changes are- Hotspot JVM included, RMI modified to support compatibility with CORBA, Java Naming and Directory Interface(JNDI) included and Java Platform Debugger Architecture(JPDA),Java Sound etc.
J2SE 1.4	6 Feb 2002	Codename merlin. This was the first release of the Java platform developed under Java community process as JSR 59. The Major Changes are- assert Keyword, Regular Expression, Exception chaining, Image I/O API for reading and writing images, Java Web start included.

J2SE 5.0	30 Sept 2004	codename Tiger , Originally Numbered 1.5, New language features added are- Generics, Metadata, Autoboxing, Enumeratations, Swing, Var args, collections static imports etc.
Java SE 6	11 Dec 2006	Codename Mustang. New changes are- Scripting Language support, Improved web service support trough JAX-WS , JDBC 4.0 support, support for pluggable annotations, JVM improvements including Synchronization and compiler performance optimizations, Garbaze collection algorithms.
Java SE 7.0	28 July 2011	Codename Dolphin, currently in development stage, The features changes are- JVM support for dynamic language, New library for parallel computing on multi core, compressed 64 bit pointers, Automatic resource management.
Java SE 8.0	18 March 2014	

2.3 Java Program Structure

Basic Structure of Java Program

- Documentation Section
- Package Statement
- Import Statements
- Interface Statement
- Class Definition
- Main Method Class
- Main Method Definition

Documentation Section	It comprises of a comment line which gives the names program, the programmer's name and some other brief details. In addition to the 2 other styles of comments i.e. /**/ **and** //, Java also provides another style of comment **i.e.** /**....*/
Package statement	The first statement allowed in Java file is the Package statement which is used to declare a package name and it informs the compiler that the classes defined within the program belong to this package. It is declared as: **package package_name;**

Import statements	The next is the number of import statements, which is equivalent to #include statement in C++. **Example:** `import calc.add;`
Interface statement	Interfaces are like class that includes a group of method declarations. This is an optional section and can be used only when programmers want to implement multiple inheritance(s) within a program.
Class Definition	A Java program may contain multiple class definitions. Classes are the main and important elements of any Java program. These classes are used to plot the objects of real world problem.
Main Method Class	Since every Java stand alone program requires a main method as the starting point of the program. This class is essentially a part of Java program. A simple Java program contains only this part of the program.

Sample Code of Java "Hello Java" Program

There may be many classes in the Java program, and only one class defines the main method. Methods contain data type declaration and executable statements.

To understand the class structure and features I have given an example of Hello Java program here. There are a few lines in the program and the major task of the program is to print ***Hello Java*** text on the screen.

Example:

```
public class Hello
{
 /* Author: www.w3schools.in
   Date: 2017-08-11
   Description:
   Writes the words "Hello Java" on the screen */

   public static void main(String[] args)
   {
      System.out.println("Hello Java");
   }
}
```

Program Output:

```
Hello Java
```

2.4 Differences between C++ vs Java

There are many differences and similarities between C++ programming language and Java. A list of top differences between C++ and Java are given below:

Comparison Index	C++	JAVA
Platform-independent	C++ is platform-dependent.	Java is platform-independent.
Mainly used for	C++ is mainly used for system programming.	Java is mainly used for application programming. It is widely used in window, web-based, enterprise and mobile applications.
Goto	C++ supports goto statement.	Java doesn't support goto statement.
Multiple inheritance	C++ supports multiple inheritance.	Java doesn't support multiple inheritance through class. It can be achieved by interfaces in java.
Operator Overloading	C++ supports operator overloading.	Java doesn't support operator overloading.
Pointers	C++ supports pointers. You can write pointer program in C++.	Java supports pointer internally. But you can't write the pointer program in java. It means java has restricted pointer support in java.
Compiler and Interpreter	C++ uses compiler only.	Java uses compiler and interpreter both.
Call by Value and Call by reference	C++ supports both call by value and call by reference.	Java supports call by value only. There is no call by reference in java.
Structure and Union	C++ supports structures and unions.	Java doesn't support structures and unions.
Thread Support	C++ doesn't have built-in support for threads. It relies on third-party libraries for thread support.	Java has built-in thread support.

Documentation comment	C++ doesn't support documentation comment.	Java supports documentation comment (/**...*/) to create documentation for java source code.
Virtual Keyword	C++ supports virtual keyword so that we can decide whether or not override a function.	Java has no virtual keyword. We can override all non-static methods by default. In other words, non-static methods are virtual by default.
unsigned right shift >>>	C++ doesn't support >>> operator.	Java supports unsigned right shift >>> operator that fills zero at the top for the negative numbers. For positive numbers, it works same like >> operator.
Inheritance Tree	C++ creates a new inheritance tree always.	Java uses single inheritance tree always because all classes are the child of Object class in java. Object class is the root of inheritance tree in java.

CHAPTER-3

JAVA – BASIC SYNTAX

When we consider a Java program, it can be defined as a collection of objects that communicate via invoking each other's methods. Let us now briefly look into what do class, object, methods, and instance variables mean.

- **Object -** Objects have states and behaviors. Example: A dog has states - color, name, breed as well as behavior such as wagging their tail, barking, eating. An object is an instance of a class.
- **Class -** A class can be defined as a template/blueprint that describes the behavior/state that the object of its type supports.
- **Methods -** A method is basically a behavior. A class can contain many methods. It is in methods where the logics are written, data is manipulated and all the actions are executed.
- **Instance Variables -** Each object has its unique set of instance variables. An object's state is created by the values assigned to these instance variables.

3.1 First Java Program

Let us look at a simple code that will print the words **Hello World.**

```
public class MyFirstJavaProgram {
/* This is my first java program.
* This will print 'Hello World' as the output
*/
public static void main(String []args) {
System.out.println("Hello World"); // prints Hello World
}
}
```

Let's look at how to save the file, compile, and run the program. Please follow the subsequent steps:

- Open notepad and add the code as above.
- Save the file as: MyFirstJavaProgram.java.
- Open a command prompt window and go to the directory where you saved the class. Assume it's C:\.
- Type 'javac MyFirstJavaProgram.java' and press enter to compile your code. If there are no errors in your code, the command prompt will take you to the next line (Assumption: The path variable is set).

- Now, type ' java MyFirstJavaProgram ' to run your program.
- You will be able to see ' Hello World ' printed on the window.

```
C:\> javac MyFirstJavaProgram.java
C:\> java MyFirstJavaProgram
Hello World
```

3.2 Basic Syntax

About Java programs, it is very important to keep in mind the following points.

- **Case Sensitivity -** Java is case sensitive, which means identifier Hello and hello would have different meaning in Java.
- **Class Names -** For all class names the first letter should be in Upper Case. If several words are used to form a name of the class, each inner word's first letter should be in Upper Case.

Example: class MyFirstJavaClass

- **Method Names -** All method names should start with a Lower Case letter. If several words are used to form the name of the method, then each inner word's first letter should be in Upper Case.

Example: public void myMethodName()

- **Program File Name -** Name of the program file should exactly match the class name. When saving the file, you should save it using the class name (Remember Java is case sensitive) and append '.java' to the end of the name (if the file name and the class name do not match, your program will not compile).

Example: Assume 'MyFirstJavaProgram' is the class name. Then the file should be saved as 'MyFirstJavaProgram.java'

- **public static void main(String args[])** - Java program processing starts from the main() method which is a mandatory part of every Java program.

3.3 Java Identifiers

All Java components require names. Names used for classes, variables, and methods are called identifiers.

In Java, there are several points to remember about identifiers. They are as follows:

- All identifiers should begin with a letter **(A to Z or a to z)**, currency character (**$**) or an underscore (_).
- After the first character, identifiers can have any combination of characters.
- A key word cannot be used as an identifier.
- Most importantly, identifiers are case sensitive.
- Examples of legal identifiers: **age, $salary, _value, __1_value.**
- Examples of illegal identifiers: **123abc, -salary.**

3.4 Java Modifiers

Like other languages, it is possible to modify classes, methods, etc., by using modifiers. There are two categories of modifiers:

- **Access Modifiers:** default, public , protected, private
- **Non-access Modifiers:** final, abstract, strictfp

3.5 Java Variables

Following are the types of variables in Java:

- Local Variables
- Class Variables (Static Variables)
- Instance Variables (Non-static Variables)

3.6 Java Arrays

Arrays are objects that store multiple variables of the same type. However, an array itself is an object on the heap. We will look into how to declare, construct, and initialize in the upcoming chapters.

3.7 Java Enums

Enums were introduced in Java 5.0. Enums restrict a variable to have one of only a few predefined values. The values in this enumerated list are called **enums.** With the use of enums it is possible to reduce the number of bugs in your code.

For example, if we consider an application for a fresh juice shop, it would be possible to restrict the glass size to small, medium, and large. This would make sure that it would not allow anyone to order any size other than small, medium, or large.

Example

```
class FreshJuice {
enum FreshJuiceSize{ SMALL, MEDIUM, LARGE }
FreshJuiceSize size;
}
public class FreshJuiceTest {
public static void main(String args[]){
FreshJuice juice = new FreshJuice();
juice.size = FreshJuice.FreshJuiceSize.MEDIUM ;
System.out.println("Size: " + juice.size);
}
}
```

Note: Enums can be declared as their own or inside a class. Methods, variables, constructors can be defined inside enums as well.

3.8 Java Keywords

The following list shows the reserved words in Java. These reserved words may not be used as constant or variable or any other identifier names.

abstract	Assert	boolean	Break
Byte	Case	catch	Char
Class	Const	continue	Default
Do	Double	else	Enum
For	Goto	if	Float
import	instanceof	int	interface
long	Native	new	Package
Private	Public	protected	Return
Short	Static	switch	This
Void	While	try	Super

3.9 Comments in Java

Java supports single-line and multi-line comments very similar to C and C++. All characters available inside any comment are ignored by Java compiler.

```
public class MyFirstJavaProgram{
/* This is my first java program.
* This will print 'Hello World' as the output
* This is an example of multi-line comments.
*/
public static void main(String []args){
// This is an example of single line comment
/* This is also an example of single line comment. */
System.out.println("Hello World");
}
}
```

3.10 Inheritance

In Java, classes can be derived from classes. Basically, if you need to create a new class and here is already a class that has some of the code you require, then it is possible to derive your new class from the already existing code.

This concept allows you to reuse the fields and methods of the existing class without having to rewrite the code in a new class. In this scenario, the existing class is called the superclass and the derived class is called the **subclass.**

3.11 Interfaces

In Java language, an interface can be defined as a contract between objects on how to communicate with each other. Interfaces play a vital role when it comes to the concept of inheritance.

CHAPTER- 4

JAVA – BASIC DATATYPES

Variables are nothing but reserved memory locations to store values. This means that when you create a variable you reserve some space in the memory.

Based on the data type of a variable, the operating system allocates memory and decides what can be stored in the reserved memory. Therefore, by assigning different datatypes to variables, you can store integers, decimals, or characters in these variables.

There are two data types available in Java:

- **Primitive Datatypes**
- **Reference/Object Datatypes**

4.1 Primitive Datatypes

There are eight primitive datatypes supported by Java. Primitive datatypes are predefined by the language and named by a keyword. Let us now look into the eight primitive data types in detail.

Byte:

- Byte data type is an 8-bit signed two's complement integer
- Minimum value is -128 (-2^7)
- Maximum value is 127 (inclusive)(2^7 -1)
- Default value is 0
- Byte datatype is used to save space in large arrays, mainly in place of integers, since a byte is four times smaller than an integer
- **Example: byte a = 100 , byte b = -50**

Short:

- Short datatype is a 16-bit signed two's complement integer
- Minimum value is -32,768 (-2^15)
- Maximum value is 32,767 (inclusive) (2^15 -1)
- Short datatype can also be used to save memory as byte data type. A short is 2 times smaller than an integer
- Default value is 0
- Example: short s = 10000, short r = -20000

Int:

- Int datatype is a 32-bit signed two's complement integer
- Minimum value is - 2,147,483,648 (-2^31)
- Maximum value is 2,147,483,647(inclusive) (2^31 -1)

- Integer is generally used as the default data type for integral values unless there is a concern about memory.
- The default value is 0
- Example: int a = 100000, int b = -200000

Long:

- Long datatype is a 64-bit signed two's complement integer
- Minimum value is -9,223,372,036,854,775,808 (-2^63)
- Maximum value is 9,223,372,036,854,775,807 (inclusive) (2^63 -1)
- This type is used when a wider range than int is needed
- Default value is 0L
- Example: long a = 100000L, long b = -200000L

Float:

- Float datatype is a single-precision 32-bit IEEE 754 floating point
- Float is mainly used to save memory in large arrays of floating point numbers
- Default value is 0.0f
- Float datatype is never used for precise values such as currency
- Example: float f1 = 234.5f

Double:

- double datatype is a double-precision 64-bit IEEE 754 floating point
- This datatype is generally used as the default data type for decimal values, generally the default choice
- Double datatype should never be used for precise values such as currency
- Default value is 0.0d
- Example: double d1 = 123.4

Boolean:

- boolean datatype represents one bit of information
- There are only two possible values: true and false
- This datatype is used for simple flags that track true/false conditions
- Default value is false
- Example: boolean one = true

Char:

- char datatype is a single 16-bit Unicode character
- Minimum value is '\u0000' (or 0)
- Maximum value is '\uffff' (or 65,535 inclusive)
- Char datatype is used to store any character
- Example: char letterA ='A'

4.2 Reference Datatypes

Let's have a discussion about Reference Data Types in brief.

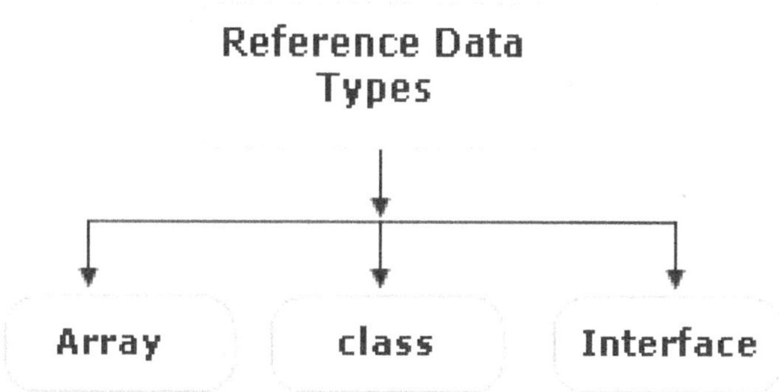

Fig: 4.1

In Java a reference data type is a variable that can contain the reference or an address of dynamically created object. These type of data type are not predefined like primitive data type. The reference data types are arrays, classes and interfaces that are made and handle according to a programmer in a java program which can hold the three kinds of values as:

array type
// Points to an array instance
class type
// Points to an object or a class instance

interface type
// Points to an object and a method, which is implemented to the corresponding interface

Class type:

As you know that Java is an object-oriented programming language where an object is a variable, associated with methods that is described by a class. The name of a class is treated as a **type** in a java program, so that you can declare a variable of an object-type, and a method which can be called using that object- type variable.

Whenever a variable is created, a reference to an object is also created using the name of a class for its type i.e. that variable can contain either **null** or a **reference** to an object of that class. It is not allowed to contain any other kinds of values. Such type is called **reference types** in Java. The object becomes an **instance** when the memory is allocated to that object using **new** keyword. In addition, **array types** are **reference types** because these are treated as objects in Java. For example:

```
class Fruit {

    fColor(){....}

    fSize(){....}

};Fruit mango;

Fruit banana;

...
```

In the given example the **Fruit** is a **class** that has the reference variables as **mango & banana** through which we can call the behaviors associated with that class as **mango.fColor();** within the main method of the super class.

Array Type:

An array is a special kind of **object** that contains values called **elements.** The java array enables the user to store the values of the same type in **contiguous** memory allocations. The elements in an array are identified by an**integer index** which initially starts from **0** and ends with **one less than number** of elements available in the array. All elements of an array must contain the same type of value i.e. if an array is a type of integer then all the elements must be of integer type. It is a **reference data type** because the class named as **Array** implicitly extends**java.lang.Object.** The syntax of declaring the array is shown as:

```
DataType [] variable1, variable2, .......variableN;

DataType [] variable = new DataType [ArraySize];


DataType [] variable = {item 1, item 2,...item n};
```

For example:

```
int [] a = new int [10];

String [] b = {"reference","data", "type"};
```

In the first statement, an array variable **"a"** is declared of integer data type that holds the memory spaces according to the size of int. The index of the array starts from **a[0]** and ends with **a[9]**. Thus, the integer value can be assigned for each or a particular **index position** of array.

In the second statement, the array **"b"** is declared of string data type that has the enough memory spaces to directly hold the three string values. Thus each value is assigned for each **index** position of the array.

Interface Type:

Java provides an another kind of **reference data type** or a mechanism to support **multiple inheritance** feature called an **interface.** The name of an interface can be used to specify the type of a reference. A value is not allowed to be assign to a variable declared using an interface type until the **object** implements the specified **interface.**

When a class declaration implements an interface, that class inherits all of the variables and methods declared in that interface. So the implementations for all of the methods declared in the interface must be provided by that class. For example, Java provides an interface called **ActionListener** whose method named **actionPerformed()** is used to handle the different kind of event. Java also provides a class called **thread** that implements **Runnable** interface. Thus the following assignment can be allowed:

```
Runnable r;

r = new Thread();
```

CHAPTER-5

JAVA – VARIABLE TYPES

A variable provides us with named storage that our programs can manipulate. Each variable in Java has a specific type, which determines the size and layout of the variable's memory; the range of values that can be stored within that memory; and the set of operations that can be applied to the variable.

You must declare all variables before they can be used. Following is the basic form of a variable declaration:

```
data type variable [ = value][, variable [= value] ...] ;
```

Here data type is one of Java's datatypes and variable is the name of the variable. To declare more than one variable of the specified type, you can use a comma-separated list.

Following are valid examples of variable declaration and initialization in Java:

```
int a, b, c; // Declares three ints, a, b, and c.
int a = 10, b = 10; // Example of initialization
byte B = 22; // initializes a byte type variable B.
double pi = 3.14159; // declares and assigns a value of PI.
char a = 'a'; // the char variable a iis initialized with value 'a'
```

This chapter will explain various variable types available in Java Language. There are three kinds of variables in Java:

- Local variables
- Instance variables
- Class/Static variables

5.1 Local Variables

- Local variables are declared in methods, constructors, or blocks.
- Local variables are created when the method, constructor or block is entered and the variable will be destroyed once it exits the method, constructor, or block.
- Access modifiers cannot be used for local variables.
- Local variables are visible only within the declared method, constructor, or block.
- Local variables are implemented at stack level internally.
- There is no default value for local variables, so local variables should be declared and an initial value should be assigned before the first use.

Example

Here, age is a local variable. This is defined inside pupAge() method and its scope is limited to only this method.

```
public class Test{
public void pupAge(){
int age = 0;
age = age + 7;
System.out.println("Puppy age is : " + age);
}
public static void main(String args[]){
Test test = new Test();
test.pupAge();
}
}
```

This will produce the following result:

```
Puppy age is: 7
```

5.2 Instance Variables

- Instance variables are declared in a class, but outside a method, constructor or any block.
- When a space is allocated for an object in the heap, a slot for each instance variable value is created.
- Instance variables are created when an object is created with the use of the keyword 'new' and destroyed when the object is destroyed.
- Instance variables hold values that must be referenced by more than one method, constructor or block, or essential parts of an object's state that must be present throughout the class.
- Instance variables can be declared in class level before or after use.
- Access modifiers can be given for instance variables.
- The instance variables are visible for all methods, constructors and block in the class. Normally, it is recommended to make these variables private (access level). However, visibility for subclasses can be given for these variables with the use of access modifiers.
- Instance variables have default values. For numbers, the default value is 0, for Booleans it is false, and for object references it is null. Values can be assigned during the declaration or within the constructor.
- Instance variables can be accessed directly by calling the variable name inside the class. However, within static methods (when instance variables are given accessibility), they should be called using the fully qualified name.

Object Reference. VariableName.
Example

```
import java.io.*;
public class Employee{
// this instance variable is visible for any child class.
public String name;
// salary variable is visible in Employee class only.
private double salary;
// The name variable is assigned in the constructor.
public Employee (String empName){
name = empName;
}
// The salary variable is assigned a value.
public void setSalary(double empSal){
salary = empSal;
}
// This method prints the employee details.
public void printEmp(){
System.out.println("name : " + name );
System.out.println("salary :" + salary);
}
public static void main(String args[]){
Employee empOne = new Employee("Ransika");
empOne.setSalary(1000);
empOne.printEmp();
}
}
```

This will produce the following result:

```
name : Ransika
salary :1000.0
```

5.3 Class/static Variables

- Class variables also known as static variables are declared with the static keyword in a class, but outside a method, constructor or a block.
- There would only be one copy of each class variable per class, regardless of how many objects are created from it.
- Static variables are rarely used other than being declared as constants. Constants are variables that are declared as public/private, final, and static. Constant variables never change from their initial value.
- Static variables are stored in the static memory. It is rare to use static variables other than declared final and used as either public or private constants.
- Static variables are created when the program starts and destroyed when the program stops.
- Visibility is similar to instance variables. However, most static variables are declared public since they must be available for users of the class.
- Default values are same as instance variables. For numbers, the default value is 0; for Booleans, it is false; and for object references, it is null. Values can be assigned during the declaration or within the constructor. Additionally, values can be assigned in special static initializer blocks.
- Static variables can be accessed by calling with the class name ClassName.VariableName.
- When declaring class variables as public static final, then variable names (constants) are all in upper case. If the static variables are not public and final, the naming syntax is the same as instance and local variables.

Example

```
import java.io.*;
public class Employee{
// salary variable is a private static variable
private static double salary;
// DEPARTMENT is a constant
public static final String DEPARTMENT = "Development ";
public static void main(String args[]){
salary = 1000;
System.out.println(DEPARTMENT + "average salary:" + salary);
}
}
```

This will produce the following result:

```
Development average salary:1000
```

CHAPTER-6

JAVA – BASIC OPERATORS

Java provides a rich set of operators to manipulate variables. We can divide all the Java operators into the following groups:

- Arithmetic Operators
- Relational Operators
- Bitwise Operators
- Logical Operators
- Assignment Operators
- Misc Operators

6.1 The Arithmetic Operators

Arithmetic operators are used in mathematical expressions in the same way that they are used in algebra. The following table lists the arithmetic operators:

Assume integer variable A holds 10 and variable B holds 20, then:

Table: 6.1

S.No	Operator and Example
1	**+ (Addition)** Adds values on either side of the operator **Example:** A + B will give 30
2	**- (Subtraction)** Subtracts right-hand operand from left-hand operand **Example:** A - B will give -10
3	*** (Multiplication)** Multiplies values on either side of the operator **Example:** A * B will give 200
4	**/ (Division)** Divides left-hand operand by right-hand operand **Example:** B / A will give 2
5	**% (Modulus)** Divides left-hand operand by right-hand operand and returns remainder **Example:** B % A will give 0
6	**++ (Increment)** Increases the value of operand by 1 **Example:** B++ gives 21
7	**-- (Decrement)** Decreases the value of operand by 1 **Example:** B-- gives 19

Example

The following program is a simple example which demonstrates the arithmetic operators. Copy and paste the following Java program in Test.java file, and compile and run this program:

```
public class Test {
public static void main(String args[]) {
int a = 10;
int b = 20;
int c = 25;
int d = 25;
System.out.println("a + b = " + (a + b) );
System.out.println("a - b = " + (a - b) );
System.out.println("a * b = " + (a * b) );
System.out.println("b / a = " + (b / a) );
System.out.println("b % a = " + (b % a) );
System.out.println("c % a = " + (c % a) );
System.out.println("a++ = " + (a++) );
System.out.println("b-- = " + (a--) );
// Check the difference in d++ and ++d
System.out.println("d++ = " + (d++) );
System.out.println("++d = " + (++d) );
}
}
```

This will produce the following result:

```
a + b = 30
a - b = -10
a * b = 200
b / a = 2
b % a = 0
c % a = 5
a++ = 10
b-- = 11
d++ = 25
++d = 27
```

6.2 The Relational Operators

There are following relational operators supported by Java language. Assume variable A holds 10 and variable B holds 20, then:

Table 6.2

S.No	Operator and Description
1	**== (equal to)** Checks if the values of two operands are equal or not, if yes then condition becomes true. **Example:** (A == B) is not true.
2	**!= (not equal to)** Checks if the values of two operands are equal or not, if values are not equal then condition becomes true. **Example:** (A != B) is true.
3	**> (greater than)** Checks if the value of left operand is greater than the value of right operand, if yes then condition becomes true. **Example:** (A > B) is not true.
4	**< (less than)** Checks if the value of left operand is less than the value of right operand, if yes then condition becomes true. **Example:** (A < B) is true.
5	**>= (greater than or equal to)** Checks if the value of left operand is greater than or equal to the value of right operand, if yes then condition becomes true. **Example:** (A >= B) is not true.
6	**<= (less than or equal to)** Checks if the value of left operand is less than or equal to the value of right operand, if yes then condition becomes true. **Example:** (A <= B) is true.

Example

The following program is a simple example that demonstrates the relational operators. Copy and paste the following Java program in Test.java file and compile and run this program.

```
public class Test {
public static void main(String args[]) {
int a = 10;
int b = 20;
System.out.println("a == b = " + (a == b) );
System.out.println("a != b = " + (a != b) );
System.out.println("a > b = " + (a > b) );
System.out.println("a < b = " + (a < b) );
System.out.println("b >= a = " + (b >= a) );
System.out.println("b <= a = " + (b <= a) );
}
}
```

This will produce the following result:

```
a == b = false
a != b = true
a > b = false
a < b = true
b >= a = true
b <= a = false
```

6.3 The Bitwise Operators

Java defines several bitwise operators, which can be applied to the integer types, long, int, short, char, and byte.

Bitwise operator works on bits and performs bit-by-bit operation. Assume if a = 60 and b = 13; now in binary format they will be as follows:

```
a = 0011 1100
b = 0000 1101
-----------------
a&b = 0000 1100
a|b = 0011 1101
a^b = 0011 0001
~a = 1100 0011
```

The following table lists the bitwise operators:
Assume integer variable A holds 60 and variable B holds 13 then:

Table: 6.3

S.No	Operator and Description
1	**& (bitwise and)** Binary AND Operator copies a bit to the result if it exists in both operands. **Example:** (A & B) will give 12 which is 0000 1100
2	**| (bitwise or)** Binary OR Operator copies a bit if it exists in either operand. **Example:** (A | B) will give 61 which is 0011 1101
3	**^ (bitwise XOR)** Binary XOR Operator copies the bit if it is set in one operand but not both. **Example:** (A ^ B) will give 49 which is 0011 0001
4	**~ (bitwise compliment)** Binary Ones Complement Operator is unary and has the effect of 'flipping' bits. **Example:** (~A) will give -61 which is 1100 0011 in 2's complement form due to a signed binary number.
5	**<< (left shift)** Binary Left Shift Operator. The left operands value is moved left by the number of bits specified by the right operand. **Example:** A << 2 will give 240 which is 1111 0000
6	**>> (right shift)** Binary Right Shift Operator. The left operands value is moved right by the number of bits specified by the right operand. **Example:** A >> 2 will give 15 which is 1111
7	**>>> (zero fill right shift)** Shift right zero fill operator. The left operands value is moved right by the number of bits specified by the right operand and shifted values are filled up with zeros. **Example:** A >>>2 will give 15 which is 0000 1111

Example

The following program is a simple example that demonstrates the bitwise operators. Copy and paste the following Java program in Test.java file and compile and run this program:

```
public class Test {
public static void main(String args[]) {
int a = 60; /* 60 = 0011 1100 */
int b = 13; /* 13 = 0000 1101 */
int c = 0;
c = a & b; /* 12 = 0000 1100 */
System.out.println("a & b = " + c );
c = a | b; /* 61 = 0011 1101 */
System.out.println("a | b = " + c );
c = a ^ b; /* 49 = 0011 0001 */
System.out.println("a ^ b = " + c );
c = ~a; /*-61 = 1100 0011 */
System.out.println("~a = " + c );
c = a << 2; /* 240 = 1111 0000 */
System.out.println("a << 2 = " + c );
c = a >> 2; /* 15 = 1111 */
System.out.println("a >> 2 = " + c );
c = a >>> 2; /* 15 = 0000 1111 */
System.out.println("a >>> 2 = " + c );
}
}
```

This will produce the following result:

```
a & b = 12
a | b = 61
a ^ b = 49
~a = -61
a << 2 = 240
a >> 15
a >>> 15
```

6.4 The Logical Operators

The following table lists the logical operators:

Assume Boolean variables A holds true and variable B holds false, then:

Table: 6.4

S.No	Description
1	**&& (logical and)** Called Logical AND operator. If both the operands are non-zero, then the condition becomes true. **Example:** (A && B) is false.

2	**| | (logical or)** Called Logical OR Operator. If any of the two operands are non-zero, then the condition becomes true. **Example:** (A | | B) is true.
3	**! (logical not)** Called Logical NOT Operator. Use to reverses the logical state of its operand. If a condition is true then Logical NOT operator will make false. **Example:** !(A && B) is true.

Example

The following simple example program demonstrates the logical operators. Copy and paste the following Java program in Test.java file and compile and run this program:

```
public class Test {
public static void main(String args[]) {
boolean a = true;
boolean b = false;
System.out.println("a && b = " + (a&&b));
System.out.println("a || b = " + (a||b) );
System.out.println("!(a && b) = " + !(a && b));
}
}
```

This will produce the following result:

```
a && b = false
a || b = true
!(a && b) = true
```

6.5 The Assignment Operators

Following are the assignment operators supported by Java language:

Table: 6.5

S.No	Operator and Description
1	= Simple assignment operator. Assigns values from right side operands to left side operand. **Example:** C = A + B will assign value of A + B into C
2	+= Add AND assignment operator. It adds right operand to the left operand and assign the result to left operand. **Example:** C += A is equivalent to C = C + A

<table>
<tr><td>3</td><td>-=
Subtract AND assignment operator. It subtracts right operand from the left operand and assign the result to left operand.
Example:C -= A is equivalent to C = C – A</td></tr>
<tr><td>4</td><td>*=
Multiply AND assignment operator. It multiplies right operand with the left operand and assign the result to left operand.
Example: C *= A is equivalent to C = C * A</td></tr>
<tr><td>5</td><td>/=
Divide AND assignment operator. It divides left operand with the right operand and assign the result to left operand.
Example: C /= A is equivalent to C = C / A</td></tr>
<tr><td>6</td><td>%=
Modulus AND assignment operator. It takes modulus using two operands and assign the result to left operand.
Example: C %= A is equivalent to C = C % A</td></tr>
<tr><td>7</td><td><<=
Left shift AND assignment operator.
Example: C <<= 2 is same as C = C << 2</td></tr>
<tr><td>8</td><td>>>=
Right shift AND assignment operator
Example: C >>= 2 is same as C = C >> 2</td></tr>
<tr><td>9</td><td>&=
Bitwise AND assignment operator.
Example: C &= 2 is same as C = C & 2</td></tr>
<tr><td>10</td><td>^=
bitwise exclusive OR and assignment operator.
Example: C ^= 2 is same as C = C ^ 2</td></tr>
<tr><td>11</td><td>|=
bitwise inclusive OR and assignment operator.
Example: C |= 2 is same as C = C | 2</td></tr>
</table>

Example

The following program is a simple example that demonstrates the assignment operators. Copy and paste the following Java program in Test.java file. Compile and run this program:

```
public class Test {
public static void main(String args[]) {
int a = 10;
int b = 20;
int c = 0;
c = a + b;
System.out.println("c = a + b = " + c );
c += a ;
System.out.println("c += a = " + c );
c -= a ;
System.out.println("c -= a = " + c );
c *= a ;
System.out.println("c *= a = " + c );
a = 10;
c = 15;
c /= a ;
System.out.println("c /= a = " + c );
a = 10;
c = 15;
c %= a ;
System.out.println("c %= a = " + c );
c <<= 2 ;
System.out.println("c <<= 2 = " + c );
c >>= 2 ;
System.out.println("c >>= 2 = " + c );
c >>= 2 ;
System.out.println("c >>= a = " + c );
c &= a ;
System.out.println("c &= 2 = " + c );
c ^= a ;
System.out.println("c ^= a = " + c );
c |= a ;
System.out.println("c |= a = " + c );
}
}
```

This will produce the following result:

```
c = a + b = 30
c += a = 40
c -= a = 30
c *= a = 300
c /= a = 1
c %= a = 5
c <<= 2 = 20
c >>= 2 = 5
c >>= 2 = 1
c &= a = 0
c ^= a = 10
c |= a = 10
```

6.6 Miscellaneous Operators

There are few other operators supported by Java Language.

Conditional Operator (? :)

Conditional operator is also known as the ternary operator. This operator consists of three operands and is used to evaluate Boolean expressions. The goal of the operator is to decide, which value should be assigned to the variable. The operator is written as:

```
variable x = (expression) ? value if true : value if false
```

Following is an example:

```
public class Test {
public static void main(String args[]){
int a, b;
a = 10;
b = (a == 1) ? 20: 30;
System.out.println( "Value of b is : " + b );
b = (a == 10) ? 20: 30;
System.out.println( "Value of b is : " + b );
}
}
```

This will produce the following result:

```
Value of b is : 30
Value of b is : 20
```

Instance of Operator

This operator is used only for object reference variables. The operator checks whether the object is of a particular type (class type or interface type). instanceof operator is written as:

```
(Object reference variable) instanceof (class/interface type)
```

If the object referred by the variable on the left side of the operator passes the IS-A check for the class/interface type on the right side, then the result will be true. Following is an example:

```
public class Test {
public static void main(String args[]){
String name = "James";
// following will return true since name is type of String
boolean result = name instanceof String;
System.out.println( result );
}
}
```

This will produce the following result:

```
true
```

This operator will still return true, if the object being compared is the assignment compatible with the type on the right. Following is one more example:

```
class Vehicle {}
public class Car extends Vehicle {
public static void main(String args[]){
Vehicle a = new Car();
boolean result = a instanceof Car;
System.out.println( result );
}
}
```

This will produce the following result:

```
true
```

CHAPTER- 7

JAVA – LOOP CONTROL

There may be a situation when you need to execute a block of code several number of times. In general, statements are executed sequentially: The first statement in a function is executed first, followed by the second, and so on.

Programming languages provide various control structures that allow for more complicated execution paths.

A **loop statement** allows us to execute a statement or group of statements multiple times and following is the general form of a loop statement in most of the programming languages:

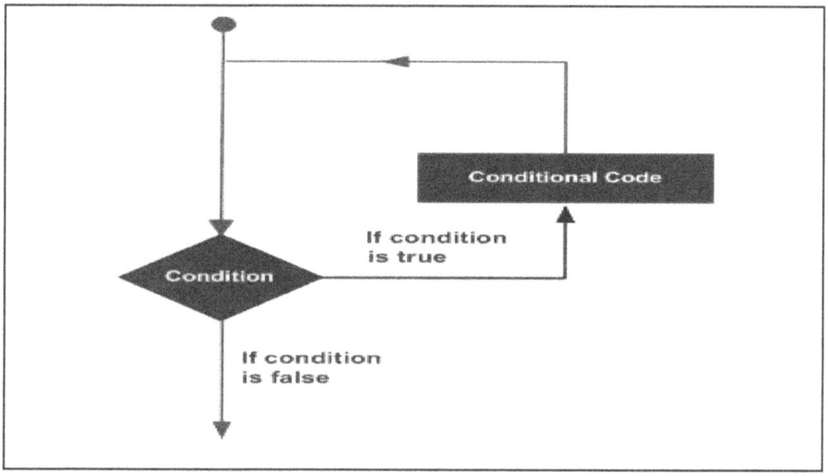

Java programming language provides the following types of loop to handle looping requirements.

Table: 7.1

Loop Type	Description
while loop	Repeats a statement or group of statements while a given condition is true. It tests the condition before executing the loop body.
for loop	Execute a sequence of statements multiple times and abbreviates the code that manages the loop variable.
do...while loop	Like a while statement, except that it tests the condition at the end of the loop body.

7.1 While Loop in Java

A while loop statement in Java programming language repeatedly executes a target statement as long as a given condition is true.

Syntax

```
while(Boolean_expression)
{
//Statements
}
```

Here, **statement(s)** may be a single statement or a block of statements. The **condition** may be any expression, and true is any non zero value.

When executing, if the **boolean_expression** result is true, then the actions inside the loop will be executed. This will continue as long as the expression result is true.

When the condition becomes false, program control passes to the line immediately following the loop.

Flow Diagram

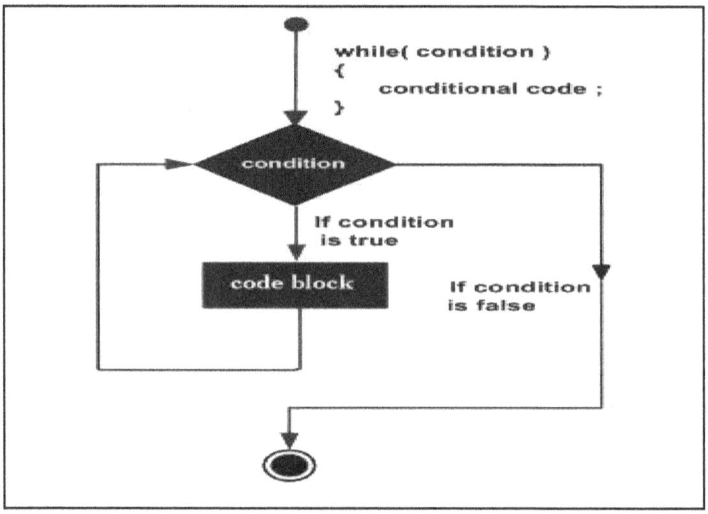

Here, key point of the while loop is that the loop might not ever run. When the expression is tested and the result is false, the loop body will be skipped and the first statement after the while loop will be executed.

Example

```
public class Test {
public static void main(String args[]) {
int x = 10;
while( x < 20 ) {
System.out.print("value of x : " + x );
x++;
System.out.print("\n");  }
}  }
```

This will produce the following result:

```
value of x : 10
value of x : 11
value of x : 12
value of x : 13
value of x : 14
value of x : 15
value of x : 16
value of x : 17
value of x : 18
value of x : 19
```

7.2 for Loop in Java

A for loop is a repetition control structure that allows you to efficiently write a loop that needs to be executed a specific number of times.

A for loop is useful when you know how many times a task is to be repeated.

Syntax

```
for(initialization; Boolean_expression; update)
{
//Statements
}
```

Here is the flow of control in a for loop:

- The **initialization** step is executed first, and only once. This step allows you to declare and initialize any loop control variables and this step ends with a **semi colon** (;).
- Next, the **Boolean expression** is evaluated. If it is true, the body of the loop is executed. If it is false, the body of the loop will not be executed and control jumps to the next statement past the for loop.

- After the **body** of the for loop gets executed, the control jumps back up to the update statement. This statement allows you to update any loop control variables. This statement can be left blank with a semicolon at the end.
- The Boolean expression is now evaluated again. If it is true, the loop executes and the process repeats (body of loop, then update step, then Boolean expression). After the Boolean expression is false, the for loop terminates.

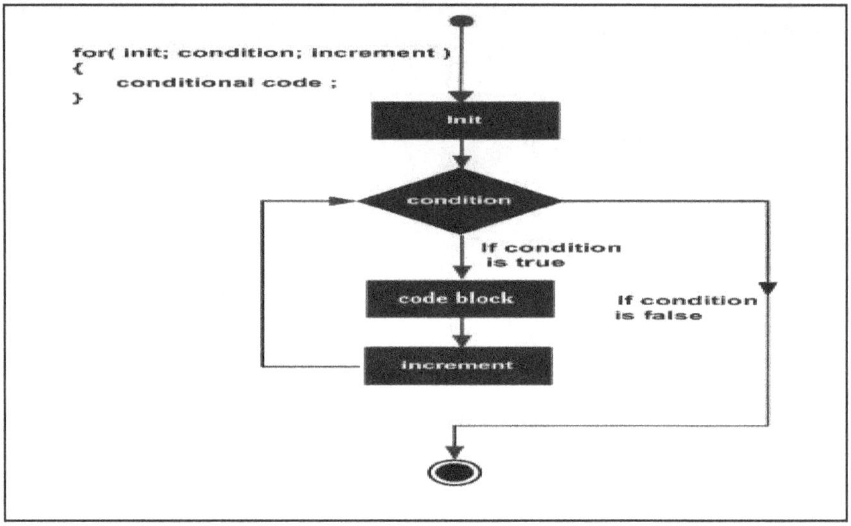

Example

Following is an example code of the for loop in Java.

```
public class Test {
public static void main(String args[]) {
for(int x = 10; x < 20; x = x+1) {
System.out.print("value of x : " + x );
System.out.print("\n");
}
}
}
```

This will produce the following result:

```
value of x : 10
value of x : 11
value of x : 12
value of x : 13
value of x : 14
value of x : 15
value of x : 16
value of x : 17
value of x : 18
value of x : 19
```

7.3 Do While Loop in Java

A do...while loop is similar to a while loop, except that a do...while loop is guaranteed to execute at least one time.

Syntax

```
do
{
//Statements
}while(Boolean_expression);
```

Notice that the Boolean expression appears at the end of the loop, so the statements in the loop execute once before the Boolean is tested.

If the Boolean expression is true, the control jumps back up to do statement, and the statements in the loop execute again. This process repeats until the Boolean expression is false.

Flow Diagram

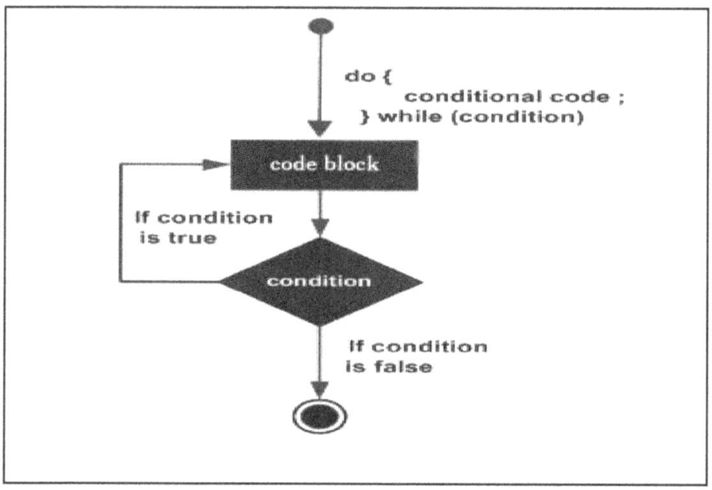

Example

```
public class Test {
public static void main(String args[]){
int x = 10;
do{
System.out.print("value of x : " + x );
x++;
System.out.print("\n");
}while( x < 20 );
}
}
```

7.4 Loop Control Statements

Loop control statements change execution from its normal sequence. When execution leaves a scope, all automatic objects that were created in that scope are destroyed.
Java supports the following control statements.

Table: 7.2

Control Statement	Description
break statement	Terminates the **loop** or **switch** statement and transfers execution to the statement immediately following the loop or switch.
continue statement	Causes the loop to skip the remainder of its body and immediately retest its condition prior to reiterating.

i) Break Statement in Java

The **break** statement in Java programming language has the following two usages:

- When the **break** statement is encountered inside a loop, the loop is immediately terminated and the program control resumes at the next statement following the loop.
- It can be used to terminate a case in the switch statement (covered in the next chapter).

Syntax

```
break;
```

Flow Diagram

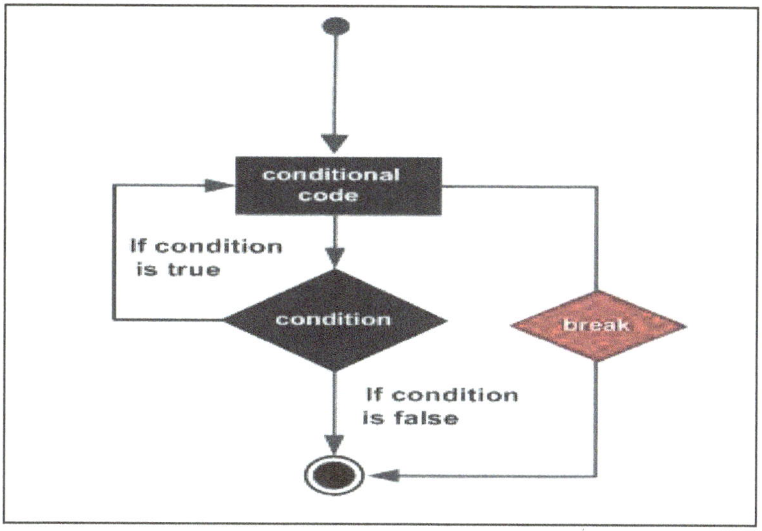

Example

```
public class Test {
public static void main(String args[]) {
int [] numbers = {10, 20, 30, 40, 50};
for(int x : numbers ) {
if( x == 30 ) {
break;
}
System.out.print( x );
System.out.print("\n");
}
}
}
```

ii) Continue Statement in Java

The **continue** keyword can be used in any of the loop control structures. It causes the loop to immediately jump to the next iteration of the loop.

- In a **for loop**, the continue keyword causes control to immediately jump to the update statement.
- In a **while loop** or do/while loop, control immediately jumps to the Boolean expression.

Syntax

```
Continue;
```

Flow Diagram

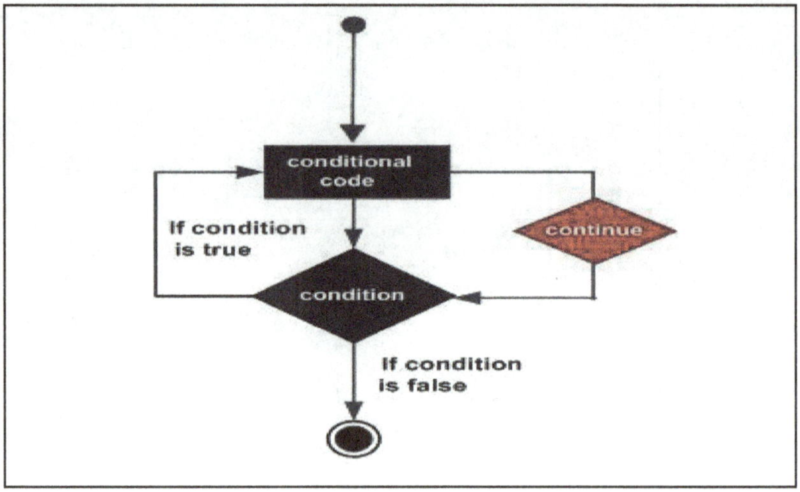

Example

```
public class Test {
public static void main(String args[]) {
int [] numbers = {10, 20, 30, 40, 50};
for(int x : numbers ) {
if( x == 30 ) {
continue;
}
System.out.print( x );
System.out.print("\n");
}
}
}
```

CHAPTER- 8

JAVA – DECISION MAKING

Decision making structures have one or more conditions to be evaluated or tested by the program, along with a statement or statements that are to be executed if the condition is determined to be true, and optionally, other statements to be executed if the condition is determined to be false.

Following is the general form of a typical decision making structure found in most of the programming languages:

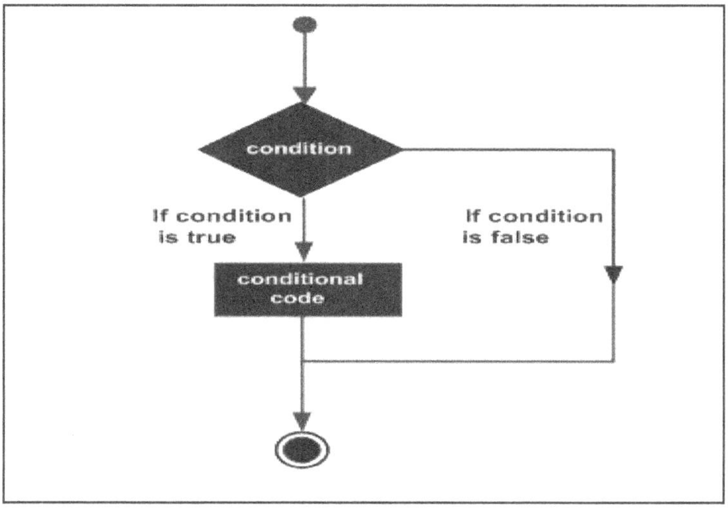

Fig: 8.1

Java programming language provides following types of decision making statements.

Table: 8.1

Statement	Description
if statement	An **if statement** consists of a boolean expression followed by one or more statements.
if...else statement	An **if statement** can be followed by an optional **else statement**, which executes when the boolean expression is false.
nested if statements	You can use one **if** or **else if** statement inside another **if** or **else if** statement(s).
switch statement	A **switch** statement allows a variable to be tested for equality against a list of values.

8.1 If Statement in Java

An **if statement** consists of a Boolean expression followed by one or more statements.

Syntax

```
if(Boolean_expression)
{
//Statements will execute if the Boolean expression is true
}
```

If the Boolean expression evaluates to true then the block of code inside the if statement will be executed. If not, the first set of code after the end of the if statement (after the closing curly brace) will be executed.

Flow Diagram

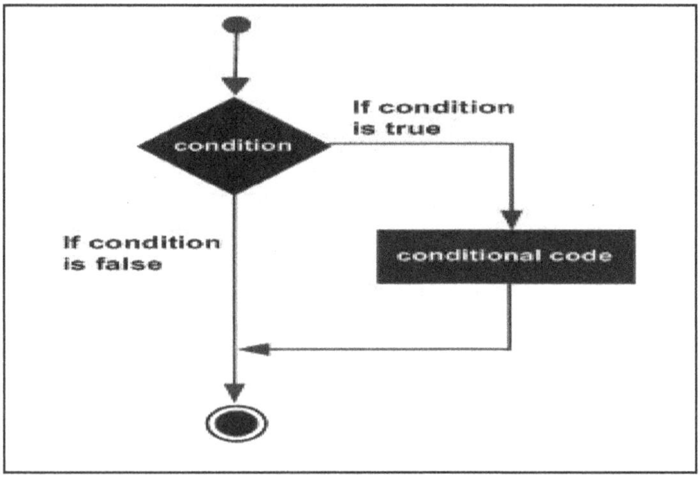

Example

```
public class Test {
public static void main(String args[]){
int x = 10;
if( x < 20 ){
System.out.print("This is if statement");   }
}
}
```

8.2 If-else Statement in Java

An if statement can be followed by an optional else statement, which executes when the Boolean expression is false.

Syntax

```
if(Boolean_expression){
//Executes when the Boolean expression is true
}else{
//Executes when the Boolean expression is false
}
```

If the boolean expression evaluates to true, then the if block of code will be executed, otherwise else block of code will be executed.

Flow Diagram

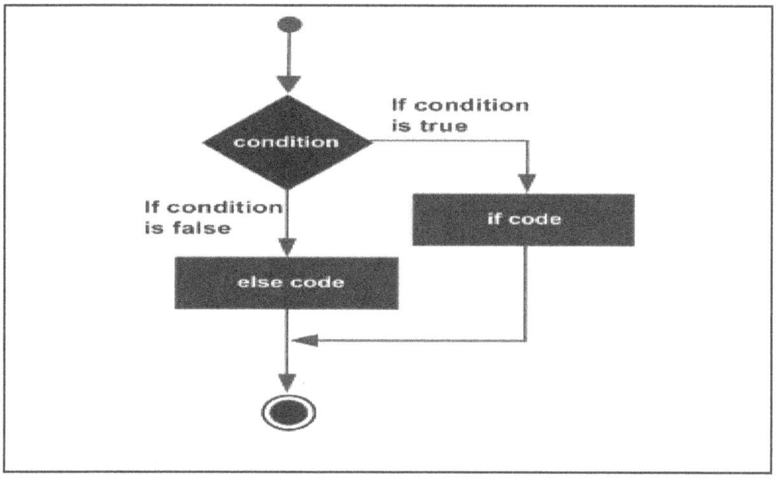

Example

```
public class Test {
public static void main(String args[]){
int x = 30;
if( x < 20 ){
System.out.print("This is if statement");
}else{
System.out.print("This is else statement");    }
}
}
```

This will produce the following result:

```
This is else statement
```

8.3 The if...else if...else Statement

An if statement can be followed by an optional else if...else statement, which is very useful to test various conditions using single if...else if statement.

When using if, else if, else statements there are a few points to keep in mind.

- An if can have zero or one else's and it must come after any else if's.
- An if can have zero to many else if's and they must come before the else.
- Once an else if succeeds, none of the remaining else if's or else's will be tested.

Syntax

```
if(Boolean_expression 1){
//Executes when the Boolean expression 1 is true
}else if(Boolean_expression 2){
//Executes when the Boolean expression 2 is true
}else if(Boolean_expression 3){
//Executes when the Boolean expression 3 is true
}else {
//Executes when the none of the above condition is true.
}
```

Example

```
public class Test {
public static void main(String args[]){
int x = 30;
if( x == 10 ){
System.out.print("Value of X is 10");
}else if( x == 20 ){
System.out.print("Value of X is 20");
}else if( x == 30 ){
System.out.print("Value of X is 30");
}else{
System.out.print("This is else statement");
}
}
}
```

This will produce the following result:

```
Value of X is 30
```

8.4 Nested if Statement in Java

It is always legal to nest if-else statements which means you can use one if or else if statement inside another if or else if statement.

Syntax

```
if(Boolean_expression 1){
//Executes when the Boolean expression 1 is true
if(Boolean_expression 2){
//Executes when the Boolean expression 2 is true
}
}
```

You can nest else if...else in the similar way as we have nested if statement.

Example

```
public class Test {
public static void main(String args[]){
int x = 30;
int y = 10;
if( x == 30 ){
if( y == 10 ){
System.out.print("X = 30 and Y = 10");
}
}
}
}
```

8.5 Switch Statement in Java

A switch statement allows a variable to be tested for equality against a list of values. Each value is called a case, and the variable being switched on is checked for each case.

Syntax

```
switch(expression){
case value :
//Statements
break; //optional
case value :
//Statements
break; //optional
//You can have any number of case statements.
default : //Optional
//Statements
}
```

The following rules apply to a switch statement:

- The variable used in a switch statement can only be integers, convertable integers (byte, short, char), strings and enums.
- You can have any number of case statements within a switch. Each case is followed by the value to be compared to and a colon.
- The value for a case must be the same data type as the variable in the switch and it must be a constant or a literal.
- When the variable being switched on is equal to a case, the statements following that case will execute until a break statement is reached.
- When a break statement is reached, the switch terminates, and the flow of control jumps to the next line following the switch statement.
- Not every case needs to contain a break. If no break appears, the flow of control will fall through to subsequent cases until a break is reached.
- A *switch* statement can have an optional default case, which must appear at the end of the switch. The default case can be used for performing a task when none of the cases is true. No break is needed in the default case.

Flow Diagram

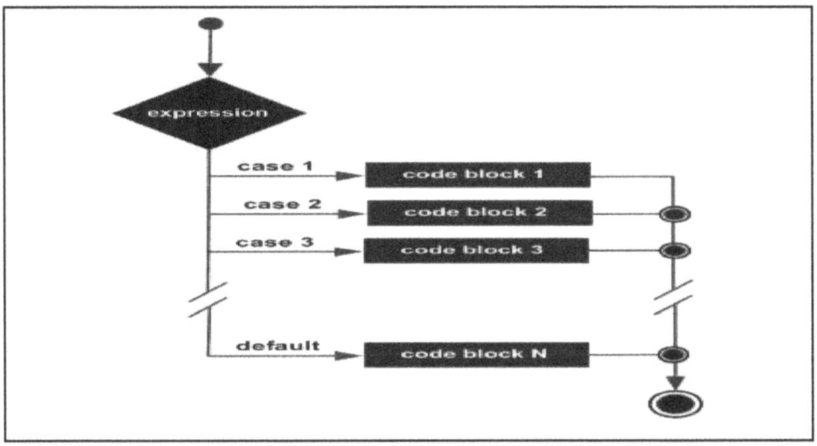

Example

```
public class Test {
public static void main(String args[]){
//char grade = args[0].charAt(0);
char grade = 'C';
switch(grade)
{
case 'A' :
System.out.println("Excellent!");
break;
case 'B' :
case 'C' :
System.out.println("Well done");
break;
case 'D' :
System.out.println("You passed");
case 'F' :
System.out.println("Better try again");
break;
default :
System.out.println("Invalid grade");
}
System.out.println("Your grade is " + grade);
}
}
```

Compile and run the above program using various command line arguments. This will produce the following result:

```
$ java Test
Well done
Your grade is a C
$
```

8.6 The? : Operator

We have covered conditional operator**?** **:** in the previous chapter which can be used to replace if...else statements. It has the following general form:

```
Exp1 ? Exp2 : Exp3;
```

Where Exp1, Exp2, and Exp3 are expressions. Notice the use and placement of the colon. To determine the value of the whole expression, initially exp1 is evaluated.

- If the value of exp1 is true, then the value of Exp2 will be the value of the whole expression.
- If the value of exp1 is false, then Exp3 is evaluated and its value becomes the value of the entire expression.

CHAPTER- 9

JAVA – STRINGS CLASS

Strings, which are widely used in Java programming, are a sequence of characters. In Java programming language, strings are treated as objects. The Java platform provides the String class to create and manipulate strings.

9.1 Creating Strings

```
String greeting = "Hello world!";
```

Whenever it encounters a string literal in your code, the compiler creates a String object with its value in this case, **"Hello world!'.**

As with any other object, you can create String objects by using the new keyword and a constructor. The String class has 11 constructors that allow you to provide the initial value of the string using different sources, such as an array of characters.

```
public class StringDemo{
public static void main(String args[]){
char[] helloArray = { 'h', 'e', 'l', 'l', 'o', '.'};
String helloString = new String(helloArray);
System.out.println( helloString );
}
}
```

9.2 Java – String Buffer & String Builder Classes:

The **StringBuffer** and **StringBuilder** classes are used when there is a necessity to make a lot of modifications to Strings of characters.

Unlike Strings, objects of type **StringBuffer** and String builder can be modified over and over again without leaving behind a lot of new unused objects.

The **StringBuilder** class was introduced as of Java 5 and the main difference between the StringBuffer and StringBuilder is that StringBuilders methods are not thread safe (not synchronised).

It is recommended to use StringBuilder whenever possible because it is faster than StringBuffer. However, if the thread safety is necessary, the best option is StringBuffer objects.

Example

```
public class Test{
public static void main(String args[]){
StringBuffer sBuffer = new StringBuffer(" test");
sBuffer.append(" String Buffer");
System.out.println(sBuffer);
}
}
```

9.3 StringBuffer Methods

Here is the list of important methods supported by StringBuffer class:

Table: 9.1

S.No	Methods with Description
1	**public StringBuffer append(String s)** Updates the value of the object that invoked the method. The method takes boolean, char, int, long, Strings, etc.
2	**public StringBuffer reverse()** The method reverses the value of the StringBuffer object that invoked the method.
3	**public delete(int start, int end)** Deletes the string starting from the start index until the end index.
4	**public insert(int offset, int i)** This method inserts a string **s** at the position mentioned by the offset.
5	**replace(int start, int end, String str)** This method replaces the characters in a substring of this StringBuffer with characters in the specified String.

9.4 String Length

Methods used to obtain information about an object are known as **accessor methods.** One accessor method that you can use with strings is the length() method, which returns the number of characters contained in the string object.

The following program is an example of **length()**, method String class.

```
public class StringDemo {
public static void main(String args[]) {
String palindrome = "Dot saw I was Tod";
int len = palindrome.length();
System.out.println( "String Length is : " + len );
}
}
```

9.5 Concatenating Strings

The String class includes a method for concatenating two strings:

```
string1.concat(string2);
```

This returns a new string that is string1 with string2 added to it at the end. You can also use the **concat**() method with string literals, as in:

```
"My name is ".concat("Zara");
```

Strings are more commonly concatenated with the **+ operator**, as in:

```
"Hello," + " world" + "!"
```

Example:

```
public class StringDemo {
public static void main(String args[]) {
String string1 = "saw I was ";
System.out.println("Dot " + string1 + "Tod");
}
}
```

9.6 Creating Format Strings

You have **printf()** and **format()** methods to print output with formatted numbers. The String class has an equivalent class method, **format()**, that returns a String object rather than a **PrintStream** object.

Using String's static **format()** method allows you to create a formatted string that you can reuse, as opposed to a one-time print statement. For example, instead of:

```
System.out.printf("The value of the float variable is " +
                  "%f, while the value of the integer " +
                  "variable is %d, and the string " +
                  "is %s", floatVar, intVar, stringVar);
```

You can write:

```
String fs;
fs = String.format("The value of the float variable is " +
                   "%f, while the value of the integer " +
                   "variable is %d, and the string " +
                   "is %s", floatVar, intVar, stringVar);
System.out.println(fs);
```

CHAPTER- 10

JAVA – ARRAYS

Java provides a data structure, the array, which stores a fixed-size sequential collection of elements of the same type. An array is used to store a collection of data, but it is often more useful to think of an array as a collection of variables of the same type.

Instead of declaring individual variables, such as number0, number1, ..., and number99, you declare one array variable such as numbers and use numbers[0], numbers[1], and ..., numbers[99] to represent individual variables.

10.1 Declaring Array Variables

To use an array in a program, you must declare a variable to reference the array, and you must specify the type of array the variable can reference. Here is the syntax for declaring an array variable

```
dataType[] arrayRefVar; // preferred way.
or
dataType arrayRefVar[]; // works but not preferred way.
```

Note: The style **dataType[] arrayRefVar** is preferred. The style **dataType arrayRefVar[]** comes from the C/C++ language and was adopted in Java to accommodate C/C++ programmers.

Example

The following code snippets are examples of this syntax:

```
double[] myList; // preferred way.
or
double myList[]; // works but not preferred way.
```

10.2 Creating Arrays

You can create an array by using the new operator with the following syntax:

```
arrayRefVar = new dataType[arraySize];
```

The above statement does two things:

- It creates an array using new **dataType[arraySize].**
- It assigns the reference of the newly created array to the variable **arrayRefVar.**

Declaring an array variable, creating an array, and assigning the reference of the array to the variable can be combined in one statement, as shown below:

```
dataType[] arrayRefVar = new dataType[arraySize];
```

Alternatively you can create arrays as follows:

```
dataType[] arrayRefVar = {value0, value1, ..., valuek};
```

The array elements are accessed through the **index.** Array indices are 0-based; that is, they start from 0 to **arrayRefVar.length-1.**

Example

Following statement declares an array variable, **myList,** creates an array of 10 elements of double type and assigns its reference to **myList:**

```
double[] myList = new double[10];
```

Following picture represents array myList. Here, myList holds ten double values and the indices are from 0 to 9.

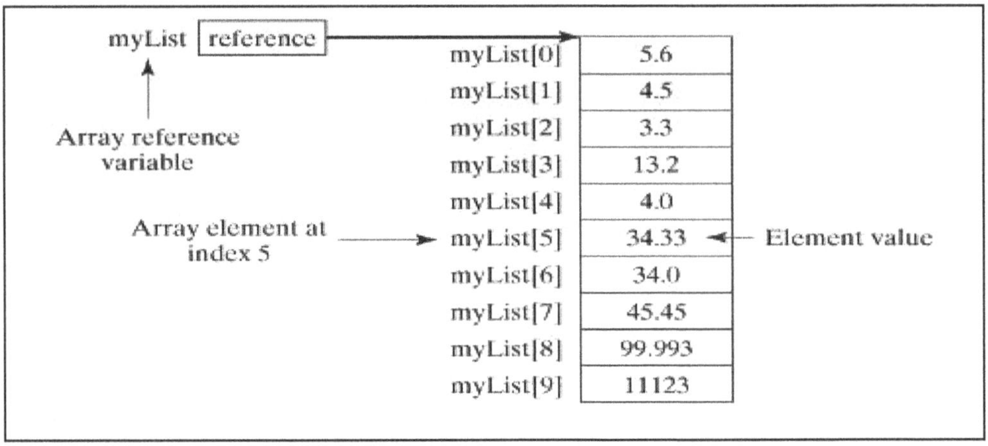

10.3 Processing Arrays

When processing array elements, we often use either for loop or foreach loop because all of the elements in an array are of the same type and the size of the array is known.

Example

Here is a complete example showing how to create, initialize, and process arrays:

```
public class TestArray {
public static void main(String[] args) {
double[] myList = {1.9, 2.9, 3.4, 3.5};
// Print all the array elements
for (int i = 0; i < myList.length; i++) {
System.out.println(myList[i] + " ");
}
// Summing all elements
double total = 0;
for (int i = 0; i < myList.length; i++) {
total += myList[i];
}
System.out.println("Total is " + total);
// Finding the largest element
double max = myList[0];
for (int i = 1; i < myList.length; i++) {
if (myList[i] > max) max = myList[i];
}
System.out.println("Max is " + max);
}
}
```

10.4 Passing Arrays to Methods

Just as you can pass primitive type values to methods, you can also pass arrays to methods. For example, the following method displays the elements in an **int** array:

```
public static void printArray(int[] array) {
for (int i = 0; i < array.length; i++) {
System.out.print(array[i] + " ");
}
}
```

You can invoke it by passing an array. For example, the following statement invokes the printArray method to display 3, 1, 2, 6, 4, and 2:

```
printArray(new int[]{3, 1, 2, 6, 4, 2});
```

10.5 Returning an Array from a Method

A method may also return an array. For example, the following method returns an array that is the reversal of another array:

```
public static int[] reverse(int[] list) {
int[] result = new int[list.length];
for (int i = 0, j = result.length - 1; i < list.length; i++, j--) {
result[j] = list[i];
}
return result;
}
```

10.6 The Arrays Class

The **java.util.Arrays** class contains various static methods for sorting and searching arrays, comparing arrays, and filling array elements. These methods are overloaded for all primitive types.

Table: 10.1

S.No	Methods with Description
1	**public static int binarySearch(Object[] a, Object key)** Searches the specified array of Object (Byte, Int , double, etc.) for the specified value using the binary search algorithm. The array must be sorted prior to making this call. This returns index of the search key, if it is contained in the list; otherwise, it returns (– (insertion point + 1)).
2	**public static boolean equals(long[] a, long[] a2)** Returns true if the two specified arrays of longs are equal to one another. Two arrays are considered equal if both arrays contain the same number of elements, and all corresponding pairs of elements in the two arrays are equal. This returns true if the two arrays are equal. Same method could be used by all other primitive data types (Byte, short, Int, etc.)
3	**public static void fill(int[] a, int val)** Assigns the specified int value to each element of the specified array of ints. The same method could be used by all other primitive data types (Byte, short, Int, etc.)
4	**public static void sort(Object[] a)** Sorts the specified array of objects into an ascending order, according to the natural ordering of its elements. The same method could be used by all other primitive data types (Byte, short, Int, etc.)

CHAPTER- 11

JAVA – DATE & TIME

Java provides the Date class available in **java.util package,** this class encapsulates the current date and time.

The Date class supports two constructors as shown in the following table.

Table: 11.1

S.No	Constructor and Description
1	**Date( )** This constructor initializes the object with the current date and time.
2	**Date(long millisec)** This constructor accepts an argument that equals the number of milliseconds that have elapsed since midnight, January 1, 1970.

Following are the methods of the date class.

Table: 11.2

S.No	Methods with Description
1	**boolean after(Date date)** Returns true if the invoking Date object contains a date that is later than the one specified by date, otherwise, it returns false.
2	**boolean before(Date date)** Returns true if the invoking Date object contains a date that is earlier than the one specified by date, otherwise, it returns false.
3	**Object clone( )** Duplicates the invoking Date object.
4	**int compareTo(Date date)** Compares the value of the invoking object with that of date. Returns 0 if the values are equal. Returns a negative value if the invoking object is earlier than date. Returns a positive value if the invoking object is later than date.
5	**int compareTo(Object obj)** Operates identically to compareTo(Date) if obj is of class Date. Otherwise, it throws a ClassCastException.
6	**boolean equals(Object date)** Returns true if the invoking Date object contains the same time and

	date as the one specified by date, otherwise, it returns false.
7	**long getTime( )** Returns the number of milliseconds that have elapsed since January 1, 1970.
8	**int hashCode( )** Returns a hash code for the invoking object.
9	**void setTime(long time)** Sets the time and date as specified by time, which represents an elapsed time in milliseconds from midnight, January 1, 1970
10	**String toString( )** Converts the invoking Date object into a string and returns the result.

11.1 Getting Current Date & Time

This is a very easy method to get current date and time in Java. You can use a simple Date object with **toString()** method to print the current date and time as follows:

```
import java.util.Date;
public class DateDemo {
public static void main(String args[]) {
// Instantiate a Date object
Date date = new Date();
// display time and date using toString()
System.out.println(date.toString());
}
}
```

This will produce the following result:

```
on May 04 09:51:52 CDT 2009
```

11.2 Date Comparison

Following are the three ways to compare two dates:

- You can use **getTime( )** to obtain the number of milliseconds that have elapsed since midnight, January 1, 1970, for both objects and then compare these two values.
- You can use the methods **before( ), after( )**, and **equals( ).** Because the 12th of the month comes before the 18th, for example, new Date(99, 2, 12).before(new Date (99, 2, 18)) returns true.
- You can use the **compareTo( )** method, which is defined by the Comparable interface and implemented by Date.

11.3 Date Formatting Using SimpleDateFormat

SimpleDateFormat is a concrete class for formatting and parsing dates in a locale-sensitive manner. **SimpleDateFormat** allows you to start by choosing any user-defined patterns for date-time formatting.

Example:

```
import java.util.*;
import java.text.*;
public class DateDemo {
public static void main(String args[]) {
Date dNow = new Date( );
SimpleDateFormat ft =
new SimpleDateFormat ("E yyyy.MM.dd 'at' hh:mm:ss a zzz");
System.out.println("Current Date: " + ft.format(dNow));
}
}
```

This will produce the following result:

```
Current Date: Sun 2004.07.18 at 04:14:09 PM PDT
```

11.4 Date formatting using printf

Date and time formatting can be done very easily using **printf** method. You use a two-letter format, starting with **t** and ending in one of the letters of the table as shown in the following code. For **example:**

```
import java.util.Date;
public class DateDemo {
public static void main(String args[]) {
// Instantiate a Date object
Date date = new Date();
// display time and date using toString()
String str = String.format("Current Date/Time : %tc", date );
System.out.printf(str);
}
}
```

This will produce the following result:

```
Current Date/Time : Sat Dec 15 16:37:57 MST 2012
```

It would be a bit silly if you had to supply the date multiple times to format each part. For that reason, a format string can indicate the index of the argument to be formatted.
The index must immediately follow the % and it must be terminated by a $. For example:

```
import java.util.Date;
public class DateDemo {
public static void main(String args[]) {
// Instantiate a Date object
Date date = new Date();
// display time and date using toString()
System.out.printf("%1$s %2$tB %2$td, %2$tY",
"Due date:", date);
}
}
```

This will produce the following result:

```
Due date: February 09, 2004
```

CHAPTER-12

JAVA – METHODS

A Java method is a collection of statements that are grouped together to perform an operation. When you call the **System.out.println()** method, for example, the system actually executes several statements in order to display a message on the console.

Now you will learn how to create your own methods with or without return values, invoke a method with or without parameters, and apply method abstraction in the program design.

12.1 Creating Method

Considering the following example to explain the syntax of a method:

```
public static int methodName(int a, int b) {
// body
}
```

Here,

- **public static:** modifier
- **int:** return type
- **methodName:** name of the method
- **a, b:** formal parameters
- **int a, int b:** list of parameters

Method definition consists of a method header and a method body. The same is shown in the following syntax:

```
modifier returnType name of Method (Parameter List) {
// method body
}
```

The syntax shown above includes:

- **modifier:** It defines the access type of the method and it is optional to use.
- **returnType:** Method may return a value.
- **nameOfMethod:** This is the method name. The method signature consists of the method name and the parameter list.

- **Parameter List:** The list of parameters, it is the type, order, and number of parameters of a method. These are optional, method may contain zero parameters.
- **method body:** The method body defines what the method does with the statements.

Example

Here is the source code of the above defined method called **max()**. This method takes two parameters num1 and num2 and returns the maximum between the two:

```
/** the snippet returns the minimum between two numbers */
public static int minFunction(int n1, int n2) {
int min;
if (n1 > n2)
min = n2;
else
min = n1;
return min;
}
```

12.2 Method Calling

For using a method, it should be called. There are two ways in which a method is called i.e., method returns a value or returning nothing (no return value).

The process of method calling is simple. When a program invokes a method, the program control gets transferred to the called **method.** This called method then returns control to the caller in two conditions, when:

1. the return statement is executed.
2. it reaches the method ending closing brace.

The methods returning void is considered as call to a statement. Lets consider an example:

```
System.out.println("This is tutorialspoint.com!");
```

The method returning value can be understood by the following example:

```
int result = sum(6, 9);
```

Example

Following is the example to demonstrate how to define a method and how to call it:

```
public class ExampleMinNumber{
public static void main(String[] args) {
int a = 11;
int b = 6;
int c = minFunction(a, b);
System.out.println("Minimum Value = " + c);
}
/** returns the minimum of two numbers */
public static int minFunction(int n1, int n2) {
int min;
if (n1 > n2)
min = n2;
else
min = n1;
return min;
}
}
```

This will produce the following result:

```
Minimum value = 6
```

12.3 The void Keyword

The void keyword allows us to create methods which do not return a value. Here, in the following example we're considering a void method **methodRankPoints.** This method is a void method, which does not return any value. Call to a void method must be a statement i.e. methodRankPoints (255.7);. It is a Java statement which ends with a semicolon as shown in the following example.

Example

```
public class ExampleVoid {
public static void main(String[] args) {
methodRankPoints(255.7);
}
public static void methodRankPoints(double points) {
if (points >= 202.5) {
System.out.println("Rank:A1");
}
else if (points >= 122.4) {
System.out.println("Rank:A2");
}
else {
System.out.println("Rank:A3");
}
}
}
```

This will produce the following result:

```
Rank:A1
```

12.4 Passing Parameters by Value

While working under calling process, an argument is to be passed. These should be in the same order as their respective parameters in the method specification. Parameters can be passed by value or by reference. Passing Parameters by Value means calling a method with a parameter. Through this, the argument value is passed to the parameter.

Example

The following program shows an example of passing parameter by value. The value of the arguments remains the same even after the method invocation.

```
public class swappingExample {
public static void main(String[] args) {
int a = 30;
int b = 45;
System.out.println("Before swapping, a = " +
a + " and b = " + b);
// Invoke the swap method
swapFunction(a, b);
System.out.println("\n**Now, Before and After swapping values will be same
here**:");
System.out.println("After swapping, a = " +
a + " and b is " + b);
}
public static void swapFunction(int a, int b) {
System.out.println("Before swapping(Inside), a = " + a
+ " b = " + b);
// Swap n1 with n2
int c = a;
a = b;
b = c;
System.out.println("After swapping(Inside), a = " + a
+ " b = " + b);
}
}
```

This will produce the following result:

```
Before swapping, a = 30 and b = 45
Before swapping (Inside), a = 30 b = 45
After swapping (Inside), a = 45 b = 30
**Now, before and After swapping values will be same here**:
After swapping, a = 30 and b is 45
```

12.5 Method Overloading

When a class has two or more methods by the same name but different parameters, it is known as method overloading. It is different from overriding. In overriding, a method has the same method name, type, number of parameters, etc.

Let's consider the example discussed earlier for finding minimum numbers of integer type. If, let's say we want to find the minimum number of double type. Then the concept of overloading will be introduced to create two or more methods with the same name but different parameters.

The following example explains the same:

```
public class ExampleOverloading{
public static void main(String[] args) {
int a = 11;
int b = 6;
double c = 7.3;
double d = 9.4;
int result1 = minFunction(a, b);
// same function name with different parameters
double result2 = minFunction(c, d);
System.out.println("Minimum Value = " + result1);
System.out.println("Minimum Value = " + result2);
}
// for integer
public static int minFunction(int n1, int n2) {
int min;
if (n1 > n2)
min = n2;
else

min = n1;
return min;
}
// for double
public static double minFunction(double n1, double n2) {
double min;
if (n1 > n2)
min = n2;
else
min = n1;
return min;
}
}
```

Overloading methods makes program readable. Here, two methods are given by the same name but with different parameters. The minimum number from integer and double types is the result.

12.6 Using Command-Line Arguments

Sometimes you will want to pass some information into a program when you run it. This is accomplished by passing command-line arguments to **main( ).**

A command-line argument is the information that directly follows the program's name on the command line when it is executed. To access the command-line arguments inside a Java program is quite easy. They are stored as strings in the String array passed to **main( ).**

Example

The following program displays all of the command-line arguments that it is called with:

```
public class CommandLine {
public static void main(String args[]){
for(int i=0; i<args.length; i++){
System.out.println("args[" + i + "]: " +
args[i]);
}
}
}
```

Try executing this program as shown here:

```
$java CommandLine this is a command line 200 -100
```

This will produce the following result:

```
args[0]: this
args[1]: is
args[2]: a
args[3]: command
args[4]: line
args[5]: 200
args[6]: -100
```

12.7 The Constructors

A constructor initializes an object when it is created. It has the same name as its class and is syntactically similar to a method. However, constructors have no explicit return type.

Typically, you will use a constructor to give initial values to the instance variables defined by the class, or to perform any other startup procedures required to create a fully formed object.

All classes have constructors, whether you define one or not, because Java automatically provides a default constructor that initializes all member variables to zero. However, once you define your own constructor, the default constructor is no longer used.

Example

Here is a simple example that uses a constructor without parameters:

```
// A simple constructor.
class MyClass {
int x;
// Following is the constructor
MyClass() {
x = 10;
}
}
```

You will have to call constructor to initialize objects as follows:

```
public class ConsDemo {
public static void main(String args[]) {
MyClass t1 = new MyClass();
MyClass t2 = new MyClass();
System.out.println(t1.x + " " + t2.x);
}
}
```

12.8 Parameterized Constructor

Most often, you will need a constructor that accepts one or more parameters. Parameters are added to a constructor in the same way that they are added to a method; just declare them inside the parentheses after the constructor's name.

Example

Here is a simple example that uses a constructor with a parameter:

```
// A simple constructor.
class MyClass {
int x;
// Following is the constructor
MyClass(int i ) {
x = i;
}
}
```

You will need to call a constructor to initialize objects as follows:

```
public class ConsDemo {
public static void main(String args[]) {
MyClass t1 = new MyClass( 10 );
MyClass t2 = new MyClass( 20 );
System.out.println(t1.x + " " + t2.x);
}
}
```

This will produce the following result:

```
This will produce the following result:
```

12.9 The this keyword

this is a keyword in Java which is used as a reference to the object of the current class, with in an instance method or a constructor. Using **this** you can refer the members of a class such as constructors, variables and methods.

Note: The keyword **this** is used only within instance methods or constructors

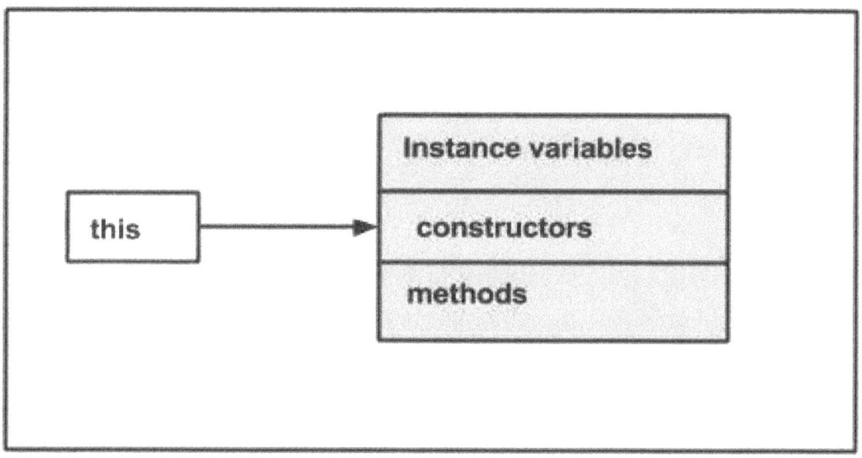

Fig: 12.1

In general, the keyword this is used to:

- Differentiate the instance variables from local variables if they have same names, within a constructor or a method.

```
class Student{
int age;
Student(int age){
this.age=age;
}
}
```

Call one type of constructor (parametrized constructor or default) from other in a class. It is known as explicit constructor invocation.

```
class Student{
int age
Student(){
this(20);
}
Student(int age){
this.age=age;
}
}
```

Example

Here is an example that uses this keyword to access the members of a class. Copy and paste the following program in a file with the name, **This_Example.java.**

```
public class This_Example {
//Instance variable num
int num=10;
This_Example(){
System.out.println("This is an example program on keyword this ");
}
This_Example(int num){
//Invoking the default constructor
this();
//Assigning the local variable num to the instance variable num
this.num=num;
}
public void greet(){
System.out.println("Hi Welcome to Tutorialspoint");
}
public void print(){
//Local variable num
int num=20;
//Printing the instance variable
System.out.println("value of local variable num is : "+num);
//Printing the local variable
System.out.println("value of instance variable num is : "+this.num);
//Invoking the greet method of a class
this.greet();
}
```

Conti.....

```
public static void main(String[] args){
//Instantiating the class
This_Example obj1=new This_Example();
//Invoking the print method
obj1.print();
//Passing a new value to the num variable through parametrized constructor
This_Example obj2=new This_Example(30);
//Invoking the print method again
obj2.print();
}
}
```

This will produce the following result:

```
This is an example program on keyword this
value of local variable num is : 20
value of instance variable num is : 10
Hi Welcome to Tutorialspoint
This is an example program on keyword this
value of local variable num is : 20
value of instance variable num is : 30
Hi Welcome to Tutorialspoint
```

12.10 Variable Arguments (var-args)

You to pass a variable number of arguments of the same type to a method. The parameter in the method is declared as follows:

```
typeName... parameterName
```

In the method declaration, you specify the type followed by an **ellipsis (...)**. Only one variable-length parameter may be specified in a method, and this parameter must be the last parameter. Any regular parameters must precede it.

Example

```
public class VarargsDemo {
public static void main(String args[]) {
// Call method with variable args
printMax(34, 3, 3, 2, 56.5);
printMax(new double[]{1, 2, 3});
}
public static void printMax( double... numbers) {
if (numbers.length == 0) {
System.out.println("No argument passed");
return;
}
double result = numbers[0];
for (int i = 1; i < numbers.length; i++)
if (numbers[i] > result)
result = numbers[i];
Java
243
System.out.println("The max value is " + result);
}
}
```

This will produce the following result:

```
The max value is 56.5

The max value is 3.0
```

12.11 The finalize( ) Method

It is possible to define a method that will be called just before an object's final destruction by the garbage collector. This method is called **finalize( )**, and it can be used to ensure that an object terminates cleanly.

For example, you might use finalize( ) to make sure that an open file owned by that object is closed.

To add a finalizer to a class, you simply define the finalize( ) method. The Java runtime calls that method whenever it is about to recycle an object of that class.

Inside the finalize( ) method, you will specify those actions that must be performed before an object is destroyed.

The finalize( ) method has this general form:

```
protected void finalize( )
{
// finalization code here
}
```

Here, the keyword protected is a specifier that prevents access to finalize( ) by code defined outside its class.

This means that you cannot know when or even if finalize( ) will be executed. For example, if your program ends before garbage collection occurs, finalize( ) will not execute.

CHAPTER-13

JAVA – EXCEPTIONS

An exception (or exceptional event) is a problem that arises during the execution of a program. When an Exception occurs the normal flow of the program is disrupted and the program/Application terminates abnormally, which is not recommended, therefore, these exceptions are to be handled.

An exception can occur for many different reasons. Following are some scenarios where an exception occurs.

- A user has entered an invalid data.
- A file that needs to be opened cannot be found.
- A network connection has been lost in the middle of communications or the JVM has run out of memory.

Some of these exceptions are caused by user error, others by programmer error, and others by physical resources that have failed in some manner.

Based on these, we have three categories of Exceptions. You need to understand them to know how exception handling works in Java.

- **Checked exceptions:** A checked exception is an exception that occurs at the compile time, these are also called as compile time exceptions. These exceptions cannot simply be ignored at the time of compilation, the programmer should take care of (handle) these exceptions.

For example, if you use FileReader class in your program to read data from a file, if the file specified in its constructor doesn't exist, then a FileNotFoundException occurs, and the compiler prompts the programmer to handle the exception.

```
import java.io.File;
import java.io.FileReader;
public class FilenotFound_Demo {
public static void main(String args[]){
File file=new File("E://file.txt");
FileReader fr = new FileReader(file);
}
}
```

If you try to compile the above program, you will get the following exceptions.

```
C:\>javac FilenotFound_Demo.java
FilenotFound_Demo.java:8: error: unreported exception
FileNotFoundException; must be caught or declared to be thrown
FileReader fr = new FileReader(file);
^
1 error
```

Note: Since the methods **read()** and **close()** of **FileReader** class throws **IOException**, you can observe that the compiler notifies to handle IOException, along with FileNotFoundException.

- **Unchecked exceptions:** An unchecked exception is an exception that occurs at the time of execution. These are also called as Runtime Exceptions. These include programming bugs, such as logic errors or improper use of an API. Runtime exceptions are ignored at the time of compilation.

For example, if you have declared an array of size 5 in your program, and trying to call the 6th element of the array then an ArrayIndexOutOfBoundsExceptionexception occurs.

```
public class Unchecked_Demo {
public static void main(String args[]){
int num[]={1,2,3,4};
System.out.println(num[5]);
}
}
```

13.1 Exception Hierarchy

All exception classes are subtypes of the **java.lang.Exception** class. The exception class is a subclass of the Throwable class. Other than the exception class there is another subclass called Error which is derived from the Throwable class.

Errors are abnormal conditions that happen in case of severe failures, these are not handled by the Java programs. Errors are generated to indicate errors generated by theruntime environment. Example: JVM is out of memory. Normally, programs cannot recover from errors.

The Exception class has two main subclasses: IOException class and Runtime Exception Class.

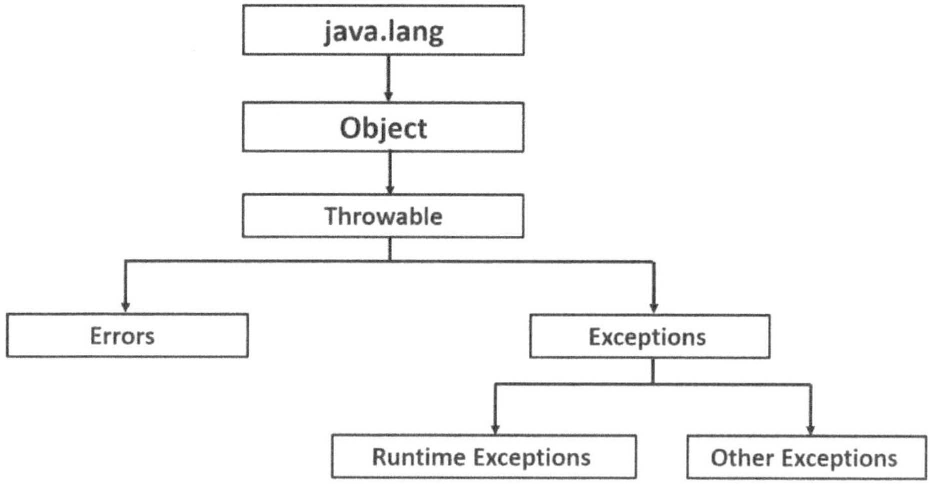

Fig: 13. 1

13.2 Exceptions Methods

Following is the list of important methods available in the Throwable class.

Table: 13.1

S.No	Methods with Description
1	public String getMessage() Returns a detailed message about the exception that has occurred. This message is initialized in the Throwable constructor.
2	public Throwable getCause() Returns the cause of the exception as represented by a Throwable object.
3	public String toString() Returns the name of the class concatenated with the result of getMessage().
4	public void printStackTrace() Prints the result of toString() along with the stack trace to System.err, the error output stream.
5	public StackTraceElement [] getStackTrace() Returns an array containing each element on the stack trace. The element at index 0 represents the top of the call stack, and the last element in the array represents the method at the bottom of the call stack.
6	public Throwable fillInStackTrace() Fills the stack trace of this Throwable object with the current stack trace, adding to any previous information in the stack trace.

13.3 Catching Exceptions

A method catches an exception using a combination of the **try and catch** keywords. A try/catch block is placed around the code that might generate an exception. Code within a try/catch block is referred to as protected code, and the syntax for using try/catch looks like the following:

```
try
{
//Protected code
}catch(ExceptionName e1)
{
//Catch block
}
```

A catch statement involves declaring the type of exception you are trying to catch. If an exception occurs in protected code, the catch block (or blocks) that follows the try is checked. If the type of exception that occurred is listed in a catch block, the exception is passed to the catch block much as an argument is passed into a method parameter.

Example

The following is an array declared with 2 elements. Then the code tries to access the 3rd element of the array which throws an exception.

```
// File Name : ExcepTest.java
import java.io.*;
public class ExcepTest{
public static void main(String args[]){
try{
int a[] = new int[2];
System.out.println("Access element three :" + a[3]);
}catch(ArrayIndexOutOfBoundsException e){
System.out.println("Exception thrown :" + e);
}
System.out.println("Out of the block");
}
}
```

13.4 Multiple Catch Blocks

A try block can be followed by multiple catch blocks. The syntax for multiple catch blocks looks like the following:

```
try
{
//Protected code
}catch(ExceptionType1 e1)
{
//Catch block
}catch(ExceptionType2 e2)
{
//Catch block
}catch(ExceptionType3 e3)
{
//Catch block
}
```

The previous statements demonstrate three catch blocks, but you can have any number of them after a single try. If an exception occurs in the protected code, the exception is thrown to the first catch block in the list. If the data type of the exception thrown matches ExceptionType1, it gets caught there. If not, the exception passes down to the second catch statement. This continues until the exception either is caught or falls through all catches, in which case the current method stops execution and the exception is thrown down to the previous method on the call stack.

Example

Here is code segment showing how to use multiple try/catch statements.

```
try
{
file = new FileInputStream(fileName);
x = (byte) file.read();
}catch(IOException i)
{
i.printStackTrace();
return -1;
}catch(FileNotFoundException f) //Not valid!
{
f.printStackTrace();
return -1;
}
```

Inheritance can be defined as the process where one class acquires the properties (methods and fields) of another. With the use of inheritance the information is made manageable in a hierarchical order.

The class which inherits the properties of other is known as subclass (derived class, child class) and the class whose properties are inherited is known as superclass (base class, parent class).

14.1 Extends Keyword

Extends is the keyword used to inherit the properties of a class. Following is the syntax of extends keyword.

```
class Super{
.....
.....
}
class Sub extends Super{
.....
.....
}
```

14.2 The super keyword

The **super** keyword is similar to **this** keyword. Following are the scenarios where the super keyword is used.

- It is used to **differentiate the members** of superclass from the members of subclass, if they have same names.
- It is used to **invoke the superclass** constructor from subclass.

14.3 Differentiating the Members

If a class is inheriting the properties of another class. And if the members of the superclass have the names same as the sub class, to differentiate these variables we use super keyword as shown below.

```
super.variable
super.method();
```

In the given program, you have two classes namely **Sub_class** and **Super_class,** both have a method named **display()** with different implementations, and a variable named num with different values. We are invoking display() method of both classes and printing the value of the variable num of both classes. Here you can observe that we have used super keyword to differentiate the members of superclass from subclass.
Copy and paste the program in a file with name **Sub_class.java**.

```
class Super_class{
int num = 20;
//display method of superclass
public void display(){
System.out.println("This is the display method of superclass");
}
}
public class Sub_class extends Super_class {
int num = 10;
//display method of sub class
public void display(){
System.out.println("This is the display method of subclass");
}
public void my_method(){

//Instantiating subclass

Sub_class sub = new Sub_class();

//Invoking the display() method of sub class

sub.display();

//Invoking the display() method of superclass

super.display();

//printing the value of variable num of subclass

System.out.println("value of the variable named num in sub class:"+
sub.num);

//printing the value of variable num of superclass

System.out.println("value of the variable named num in super class:"+
super.num);

}
```

Conti.....

```
public static void main(String args[]){
Sub_class obj = new Sub_class();
obj.my_method();
}
}
```

Compile and execute the above code using the following syntax.

```
javac Super_Demo
java Super
```

14.4 Types of Inheritance

There are various types of inheritance as demonstrated below.

Fig 14.1

Single Inheritance

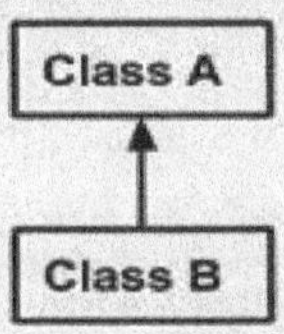


```
public class A {
  .......
}
public class B extends A {
  .........
}
```

Multi Level Inheritance

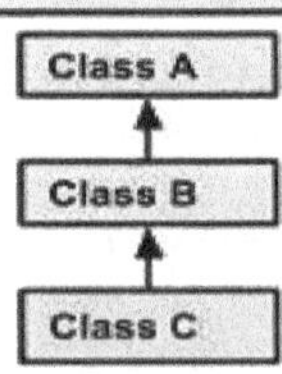


```
public class A { ...................}
public class B extends A {...................}
public class C extends  B {..................... }
```

Hierarchical Inheritance

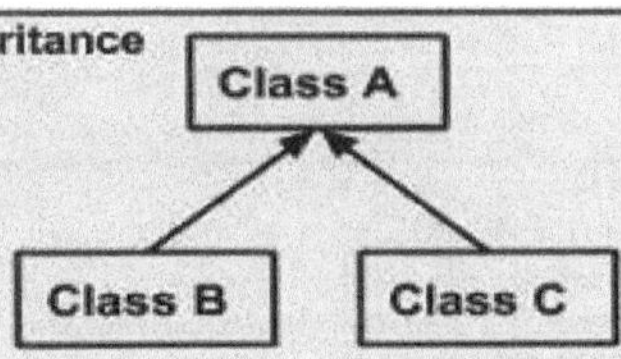


```
public class A { ...................}
public class B extends A {...................}
public class C extends A {..................... }
```

Multiple Inheritance

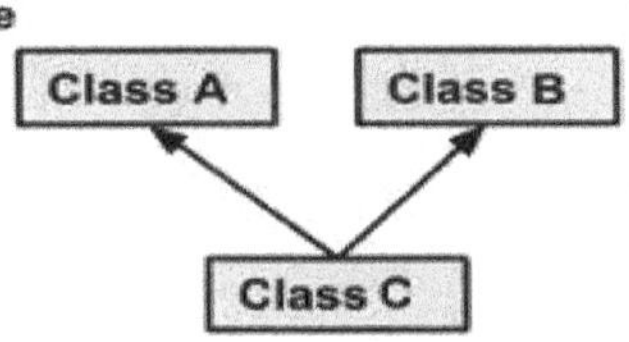


```
public class A { ...................}
public class B {...................}
public class C extends  A,B {
  .....................
} // Java does not support mutiple Inheritance
```

CHAPTER-14

JAVA – OVERRIDING

In any object-oriented programming language, Overriding is a feature that allows a subclass or child class to provide a specific implementation of a method that is already provided by one of its super-classes or parent classes. When a method in a subclass has the same name, same parameters or signature and same return type(or sub-type) as a method in its super-class, then the method in the subclass is said to override the method in the super-class.

Fig: 15.1

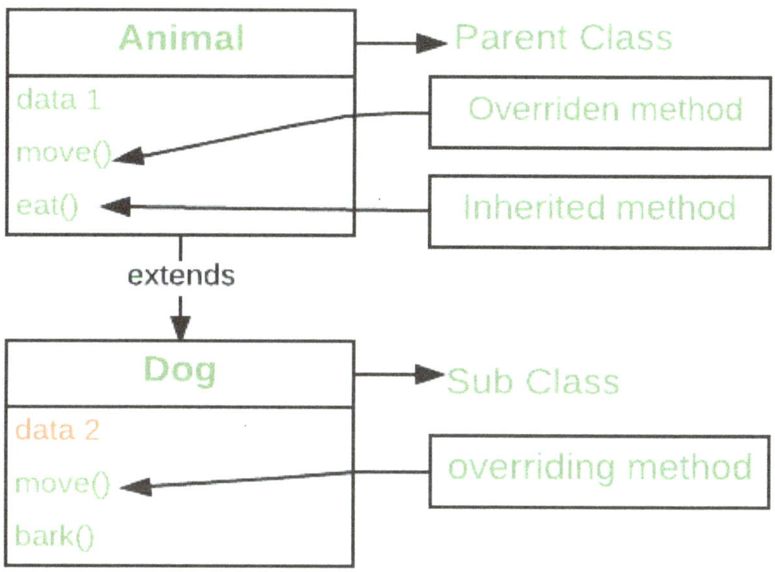

Method overriding is one of the way by which java achieve Run Time Polymorphism.The version of a method that is executed will be determined by the object that is used to invoke it. If an object of a parent class is used to invoke the method, then the version in the parent class will be executed, but if an object of the subclass is used to invoke the method, then the version in the child class will be executed.In other words, it is the type of the object being referred to (not the type of the reference variable) that determines which version of an overridden method will be executed.

Example

```
class Animal{
public void move(){
System.out.println("Animals can move");
}
}
class Dog extends Animal{
public void move(){
System.out.println("Dogs can walk and run");
}
}
public class TestDog{
public static void main(String args[]){
Animal a = new Animal(); // Animal reference and object
Animal b = new Dog(); // Animal reference but Dog object
a.move();// runs the method in Animal class
b.move();//Runs the method in Dog class
}
}
```

15.1 Rules for Method Overriding

1. Overriding and Access-Modifiers:

The access modifier for an overriding method can allow more, but not less, access than the overridden method.

For example, a protected instance method in the super-class can be made public, but not private, in the subclass. It will generate compile-time error.

```
// A Simple Java program to demonstrate
// Overriding and Access-Modifiers

class Parent
{
   // private methods are not overridden
   private void m1() { System.out.println("From parent m1()");}

   protected void m2() { System.out.println("From parent m2()"); }
}

class Child extends Parent
{
   // new m1() method
   // unique to Child class
   private void m1() { System.out.println("From child m1()");}

   // overriding method
   // with more accessibility
   @Override
   public void m2() { System.out.println("From child m2()");}

}

// Driver class
class Main
{
   public static void main(String[] args)
   {
      Parent obj1 = new Parent();
      obj1.m2();
      Parent obj2 = new Child();
      obj2.m2();
   }
}
```

Output:

```
From parent m2()
From child m2()
```

2. Final methods can not be overridden: If we don't want a method to be overridden, we declare it as final. Please see Using final with Inheritance.

```
// A Java program to demonstrate that
// final methods cannot be overridden

class Parent
{
   // Can't be overridden
   final void show() { }
}

class Child extends Parent
{
   // This would produce error
   void show() { }
}
```

3. Static methods can not be overridden (Method Overriding vs Method Hiding): When you define a static method with same signature as a static method in base class, it is known as method hiding.

The following table summarizes what happens when you define a method with the same signature as a method in a super-class.

	SUPERCLASS INSTANCE METHOD	**SUPERCLASS STATIC METHOD**
SUBCLASS INSTANCE METHOD	Overrides	Generates a compile-time error
SUBCLASS STATIC METHOD	Generates a compile-time error	Hides

```
/* Java program to show that if static method is redefined by
a derived class, then it is not overriding,it is hiding */

class Parent
{
   // Static method in base class which will be hidden in subclass
   static void m1() { System.out.println("From parent static m1()");}

   // Non-static method which will be overridden in derived class
   void m2() { System.out.println("From parent non-static(instance) m2()"); }
}

class Child extends Parent
{
   // This method hides m1() in Parent
   static void m1() { System.out.println("From child static m1()");}

   // This method overrides m2() in Parent
   @Override
   public void m2() { System.out.println("From child non-static(instance) m2()");}

}

// Driver class
class Main
{
   public static void main(String[] args)
   {
      Parent obj1 = new Child();

      // As per overriding rules this should call to class Child static
      // overridden method. Since static method can not be overridden, it
      // calls Parent's m1()
      obj1.m1();

      // Here overriding works and Child's m2() is called
      obj1.m2();
   }
```

4. Private methods can not be overridden: Private methods cannot be overridden as they are bonded during compile time. Therefore we can't even override private methods in a subclass.(See this for details).

5. The overriding method must have same return type (or subtype): From Java 5.0 onwards it is possible to have different return type for a overriding method in child class, but child's return type should be sub-type of parent's return type. This phenomenon is known as covariant return type.

6. Invoking overridden method from sub-class: We can call parent class method in overriding method using super keyword.

```
// A Java program to demonstrate that overridden
// method can be called from sub-class

// Base Class
class Parent
{
   void show()
   {
      System.out.println("Parent's show()");
   }
}

// Inherited class
class Child extends Parent
{
   // This method overrides show() of Parent
   @Override
   void show()
   {
      super.show();
      System.out.println("Child's show()");
   }
}

// Driver class
class Main
{
   public static void main(String[] args)
   {
      Parent obj = new Child();
      obj.show();
   }
}
```

7. Overriding and constructor: We can not override constructor as parent and child class can never have constructor with same name (Constructor name must always be same as Class name).

8. Overriding and Exception-Handling: Below are two rules to note when overriding methods related to exception-handling.

Rule1: If the super-class overridden method does not throws an exception, subclass overriding method can only throws the unchecked exception, throwing checked exception will lead to compile-time error.

```
/* Java program to demonstrate overriding when
superclass method does not declare an exception
*/

class Parent
{
void m1() { System.out.println("From parent m1()");}

void m2() { System.out.println("From parent  m2()"); }
}

class Child extends Parent
{
@Override
// no issue while throwing unchecked exception
void m1() throws ArithmeticException
{ System.out.println("From child m1()");}

@Override
// compile-time error
// issue while throwin checked exception
void m2() throws Exception{ System.out.println("From child m2");}

}
```

Rule2: If the super-class overridden method does throws an exception, subclass overriding method can only throw same, subclass exception. Throwing parent exception in Exception hierarchy will lead to compile time error.Also there is no issue if subclass overridden method is not throwing any exception.

```
/* Java program to demonstrate overriding when
superclass method does declare an exception
*/
class Parent
{
void m1() throws RuntimeException
{ System.out.println("From parent m1()");}
}
class Child1 extends Parent
{
@Override
// no issue while throwing same exception
void m1() throws RuntimeException
{ System.out.println("From child1 m1()");}
}
class Child2 extends Parent
{
@Override
// no issue while throwing subclass exception
void m1() throws ArithmeticException
{ System.out.println("From child2 m1()");}
}
class Child3 extends Parent
{
@Override
// no issue while not throwing any exception
void m1()
{ System.out.println("From child3 m1()");}
}
class Child4 extends Parent
{
@Override
// compile-time error
// issue while throwing parent exception
void m1() throws Exception
{ System.out.println("From child4 m1()");}

}
```

9. Overriding and abstract method: Abstract methods in an interface or abstract class are meant to be overridden in derived concrete classes otherwise compile-time error will be thrown.

10. Overriding and synchronized/stricfp method: The presence of synchronized/stricfp modifier with method has no effect on the rules of overriding, i.e. it's possible that a synchronized/stricfp method can override a non synchronized/stricfp one and vice-versa.

Why Method Overriding?

As stated earlier, overridden methods allow Java to support run-time polymorphism. Polymorphism is essential to object-oriented programming for one reason: it allows a general class to specify methods that will be common to all of its derivatives, while allowing subclasses to define the specific implementation of some or all of those methods.Overridden methods are another way that Java implements the "one interface, multiple methods" aspect of polymorphism.

Dynamic Method Dispatch is one of the most powerful mechanisms that object oriented design brings to bear on code reuse and robustness. The ability of existing code libraries to call methods on instances of new classes without recompiling while maintaining a clean abstract interface is a profoundly powerful tool.

Overridden methods allow us to call methods of any of the derived classes without even knowing type of derived class object.

CHAPTER-15

JAVA – POLYMORPHISM

This is a concept by which we can perform a single action by different ways. Polymorphism is derived from 2 Greek words: **poly and morphs.** The word "poly" means many and "morphs" means forms. So polymorphism means many forms.

There are two types of polymorphism in java: compile time polymorphism and runtime polymorphism. We can perform polymorphism in java by method overloading and method overriding.

If you overload static method in java, it is the example of compile time polymorphism. Here, we will focus on runtime polymorphism in java.

16.1 Static Polymorphism (compile time polymorphism/ Method overloading):

The ability to execute different method implementations by altering the argument used with the method name is known as method overloading. In below program, we have three print methods each with different arguments. When you properly overload a method, you can call it providing different argument lists, and the appropriate version of the method executes.

```
package oopsconcept;
class Overloadsample {
   public void print(String s){
      System.out.println("First Method with only String- "+ s);
   }
   public void print (int i){
      System.out.println("Second Method with only int- "+ i);
   }
   public void print (String s, int i){
      System.out.println("Third Method with both- "+ s + "--" + i);
   }
}
public class PolymDemo {
   public static void main(String[] args) {
      Overloadsample obj = new Overloadsample();
      obj.print(10);
      obj.print("Amit");
      obj.print("Hello", 100);
   }
}
```

16.2 Dynamic Polymorphism (run time polymorphism/ Method Overriding)

When you create a subclass by extending an existing class, the new subclass contains data and methods that were defined in the original superclass. In other words, any child class object has all the attributes of its parent. Sometimes, however, the superclass data fields and methods are not entirely appropriate for the subclass objects; in these cases, you want to override the parent class members. Let's take the example used in inheritance explanation.

```
package oopsconcept;
public class OverridingDemo {
   public static void main(String[] args) {
      //Creating Object of SuperClass and calling getModel Method
      Mobile m = new Mobile("Nokia", "Win8", "Lumia",10000);
      System.out.println(m.getModel());
      //Creating Object of Sublcass and calling getModel Method
      Android a = new Android("Samsung", "Android", "Grand",30000);
      System.out.println(a.getModel());
      //Creating Object of Sublcass and calling getModel Method
      Blackberry b = new Blackberry("BlackB", "RIM", "Curve",20000);
      System.out.println(b.getModel());
   }
}
```

Java Runtime Polymorphism Example: Bank

Consider a scenario, Bank is a class that provides method to get the rate of interest. But, rate of interest may differ according to banks. For example, SBI, ICICI and AXIS banks are providing 8.4%, 7.3% and 9.7% rate of interest.

Fig: 16.1

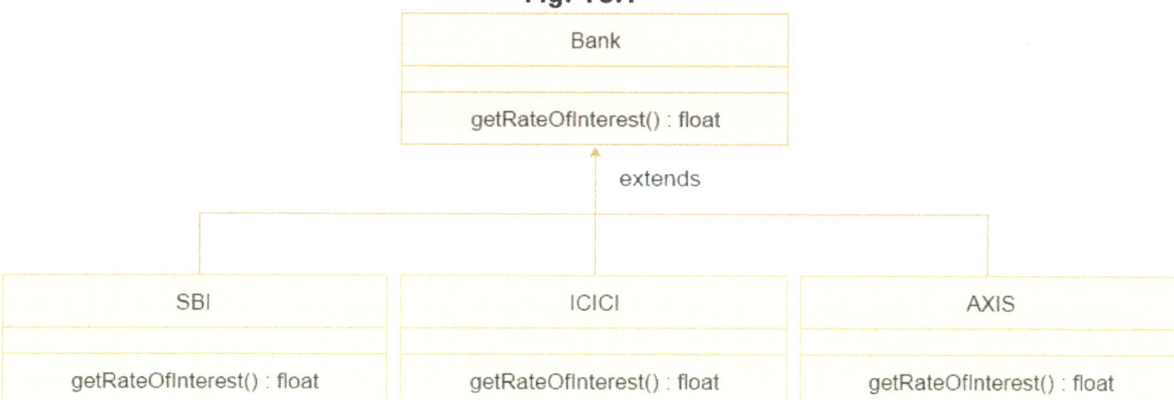

```
class Bank{
float getRateOfInterest(){return 0;}
}
class SBI extends Bank{
float getRateOfInterest(){return 8.4f;}
}
class ICICI extends Bank{
float getRateOfInterest(){return 7.3f;}
}
class AXIS extends Bank{
float getRateOfInterest(){return 9.7f;}
}
class TestPolymorphism{
public static void main(String args[]){
Bank b;
b=new SBI();
System.out.println("SBI Rate of Interest: "+b.getRateOfInterest());
b=new ICICI();
System.out.println("ICICI Rate of Interest: "+b.getRateOfInterest());
b=new AXIS();
System.out.println("AXIS Rate of Interest: "+b.getRateOfInterest());
}
}
```

Output:

```
SBI Rate of Interest: 8.4

ICICI Rate of Interest: 7.3

AXIS Rate of Interest: 9.7
```

16.3 Java Runtime Polymorphism with Data Member

Method is overridden not the datamembers, so runtime polymorphism can't be achieved by data members.

In the example given below, both the classes have a datamember speedlimit; we are accessing the datamember by the reference variable of Parent class which refers to the subclass object. Since we are accessing the datamember which is not overridden, hence it will access the datamember of Parent class always.

```
class Bike{

 int speedlimit=90;

}

class Honda3 extends Bike{

 int speedlimit=150;


 public static void main(String args[]){

  Bike obj=new Honda3();

  System.out.println(obj.speedlimit);//90

}
```

CHAPTER-16

JAVA – ABSTRACTION

As per dictionary, abstraction is the quality of dealing with ideas rather than events. For example, when you consider the case of e-mail, complex details such as what happens as soon as you send an e-mail, the protocol your e-mail server uses are hidden from the user. Therefore, to send an e-mail you just need to type the content, mention the address of the receiver, and click send.

Likewise in Object-oriented programming, abstraction is a process of hiding the implementation details from the user, only the functionality will be provided to the user. In other words, the user will have the information on what the object does instead of how it does it.

17.1 Abstract Class

A class which contains the abstract keyword in its declaration is known as abstract class.

- Abstract classes may or may not contain abstract methods, i.e., methods without body (public void get();)
- But, if a class has at least one abstract method, then the class must be declared abstract.
- If a class is declared abstract, it cannot be instantiated.
- To use an abstract class, you have to inherit it from another class, provide implementations to the abstract methods in it.
- If you inherit an abstract class, you have to provide implementations to all the abstract methods in it.

Example

This section provides you an example of the abstract class. To create an abstract class, just use the **abstract** keyword before the class keyword, in the class declaration.

```
/* File name : Employee.java */
public abstract class Employee
{
private String name;
private String address;
private int number;
public Employee(String name, String address, int number)

{
System.out.println("Constructing an Employee");
this.name = name;
this.address = address;
this.number = number;
}
public double computePay()
{
System.out.println("Inside Employee computePay");
return 0.0;
}

public void mailCheck()
{
System.out.println("Mailing a check to " + this.name
+ " " + this.address);
}

public String toString()
{
return name + " " + address + " " + number;
}
public String getName()
{
return name;
}
public String getAddress()
{
return address;
}
public void setAddress(String newAddress)
{
address = newAddress;
}
public int getNumber()
{
return number;
}
}
```

17.2 Abstract Methods

If you want a class to contain a particular method but you want the actual implementation of that method to be determined by child classes, you can declare the method in the parent class as an abstract.

- **abstract** keyword is used to declare the method as abstract.
- You have to place the **abstract** keyword before the method name in the method declaration.
- An abstract method contains a method signature, but no method body.
- Instead of curly braces, an abstract method will have a semoi colon (;) at the end.

Following is an example of the abstract method.

```
public abstract class Employee
{
private String name;
private String address;
private int number;
public abstract double computePay();
//Remainder of class definition
}
```

Declaring a method as abstract has two consequences:

- The class containing it must be declared as abstract.
- Any class inheriting the current class must either override the abstract method or declare itself as abstract.

Note: Eventually, a descendant class has to implement the abstract method; otherwise, you would have a hierarchy of abstract classes that cannot be instantiated.

CHAPTER-17

JAVA – ENCAPSULATION

Encapsulation is one of the four fundamental OOP concepts. The other three are inheritance, polymorphism, and abstraction.

Encapsulation in Java is a mechanism of wrapping the data (variables) and code acting on the data (methods) together as a single unit. In encapsulation, the variables of a class will be hidden from other classes, and can be accessed only through the methods of their current class. Therefore, it is also known as data hiding.

To achieve encapsulation in Java:

- Declare the variables of a class as private.
- Provide public setter and getter methods to modify and view the variables values.

Example

```
/* File name : EncapTest.java */
public class EncapTest{
private String name;
private String idNum;
private int age;
public int getAge(){
return age;
}
public String getName(){
return name;
}
public String getIdNum(){
return idNum;
}
public void setAge( int newAge){
age = newAge;
}
public void setName(String newName){
name = newName;
}
public void setIdNum( String newId){
idNum = newId;
}
}
```

The public setXXX() and getXXX() methods are the access points of the instance variables of the EncapTest class. Normally, these methods are referred as getters and setters. Therefore, any class that wants to access the variables should access them through these getters and setters.

Advantages of Encapsulation:

- **Data Hiding:** The user will have no idea about the inner implementation of the class. It will not be visible to the user that how the class is storing values in the variables. He only knows that we are passing the values to a setter method and variables are getting initialized with that value.
- **Increased Flexibility:** We can make the variables of the class as read-only or write-only depending on our requirement. If we wish to make the variables as read-only then we have to omit the setter methods like **setName(), setAge()** etc. from the above program or if we wish to make the variables as write-only then we have to omit the get methods like **getName(), getAge()** etc. from the above program
- **Reusability:** Encapsulation also improves the re-usability and easy to change with new requirements.
- **Testing code is easy:** Encapsulated code is easy to test for unit testing.

CHAPTER-18

JAVA – INTERFACES

An interface is a reference type in Java. It is similar to class. It is a collection of abstract methods. A class implements an interface, thereby inheriting the abstract methods of the interface.

Along with abstract methods, an interface may also contain constants, default methods, static methods, and nested types. Method bodies exist only for default methods and static methods.

Writing an interface is similar to writing a class. But a class describes the attributes and behaviors of an object. And an interface contains behaviors that a class implements.

Unless the class that implements the interface is abstract, all the methods of the interface need to be defined in the class.

An interface is similar to a class in the following ways:

- An interface can contain any number of methods.
- An interface is written in a file with a .java extension, with the name of the interface matching the name of the file.
- The byte code of an interface appears in a .class file.
- Interfaces appear in packages, and their corresponding bytecode file must be in a directory structure that matches the package name.

However, an interface is different from a class in several ways, including:

- You cannot instantiate an interface.
- An interface does not contain any constructors.
- All of the methods in an interface are abstract.
- An interface cannot contain instance fields. The only fields that can appear in an interface must be declared both static and final.
- An interface is not extended by a class; it is implemented by a class.
- An interface can extend multiple interfaces.

19.1 Declaring Interfaces

The **interface** keyword is used to declare an interface. Here is a simple example to declare an interface.

Example

```
/* File name : NameOfInterface.java */
import java.lang.*;
//Any number of import statements
public interface NameOfInterface
{
//Any number of final, static fields
//Any number of abstract method declarations\
}
```

Interfaces have the following properties:

- An interface is implicitly abstract. You do not need to use the abstract keyword while declaring an interface.
- Each method in an interface is also implicitly abstract, so the abstract keyword is not needed.
- Methods in an interface are implicitly public.

Example

```
/* File name : Animal.java */
interface Animal {
public void eat();
public void travel();
}
```

19.2 Implementing Interfaces

When a class implements an interface, you can think of the class as signing a contract, agreeing to perform the specific behaviors of the interface. If a class does not perform all the behaviors of the interface, the class must declare itself as abstract.

When overriding methods defined in interfaces, there are several rules to be followed:

- Checked exceptions should not be declared on implementation methods other than the ones declared by the interface method or subclasses of those declared by the interface method.
- The signature of the interface method and the same return type or subtype should be maintained when overriding the methods.
- An implementation class itself can be abstract and if so, interface methods need not be implemented.

When implementation interfaces, there are several rules:

- A class can implement more than one interface at a time.
- A class can extend only one class, but implement many interfaces.

- An interface can extend another interface, in a similar way as a class can extend another class.

19.3 Extending Interfaces

An interface can extend another interface in the same way that a class can extend another class. The **extends** keyword is used to extend an interface, and the child interface inherits the methods of the parent interface.

19.4 Extending Multiple Interfaces

A Java class can only extend one parent class. Multiple inheritances are not allowed. Interfaces are not classes, however, and an interface can extend more than one parent interface.

The **extends** keyword is used once, and the parent interfaces are declared in a comma-separated list.

For example, if the Hockey interface extended both Sports and Event, it would be declared as:

```
public interface Hockey extends Sports, Event
```

CHAPTER-19

JAVA – PACKAGES

Packages are used in Java in order to prevent naming conflicts, to control access, to make searching/locating and usage of classes, interfaces, enumerations and annotations easier, etc.

A Package can be defined as a grouping of related types (classes, interfaces, enumerations and annotations) providing access protection and namespace management.

Some of the existing packages in Java are:

- **java.lang** - bundles the fundamental classes
- **java.io** - classes for input, output functions are bundled in this package

Programmers can define their own packages to bundle group of classes/interfaces, etc. It is a good practice to group related classes implemented by you so that a programmer can easily determine that the classes, interfaces, enumerations, and annotations are related.

Since the package creates a new namespace there won't be any name conflicts with names in other packages. Using packages, it is easier to provide access control and it is also easier to locate the related classes.

20.1 Creating a Package

While creating a package, you should choose a name for the package and include a package statement along with that name at the top of every source file that contains the classes, interfaces, enumerations, and annotation types that you want to include in the package.

The package statement should be the first line in the source file. There can be only one package statement in each source file, and it applies to all types in the file.

If a package statement is not used then the class, interfaces, enumerations, and annotation types will be placed in the current default package.

To compile the Java programs with package statements, you have to use -d option as shown below.

```
javac -d Destination_folder file_name.java
```

Then a folder with the given package name is created in the specified destination, and the compiled class files will be placed in that folder.

Example

Let us look at an example that creates a package called animals. It is a good practice to use names of packages with lower case letters to avoid any conflicts with the names of classes and interfaces.

```
/* File name : Animal.java */
package animals;
interface Animal {
public void eat();
public void travel();
}
```

Now, let us implement the above interface in the same package animals:

```
package animals;
/* File name : MammalInt.java */
public class MammalInt implements Animal{
public void eat(){
System.out.println("Mammal eats");
}
public void travel(){
System.out.println("Mammal travels");
}
public int noOfLegs(){
return 0;
}
public static void main(String args[]){
MammalInt m = new MammalInt();
m.eat();
m.travel();
}
}
```

Now compile the java files as shown below:

```
$ javac -d . Animal.java
$ javac -d . MammalInt.java
```

Now a package/folder with the name animals will be created in the current directory and these class files will be placed in it as shown below.

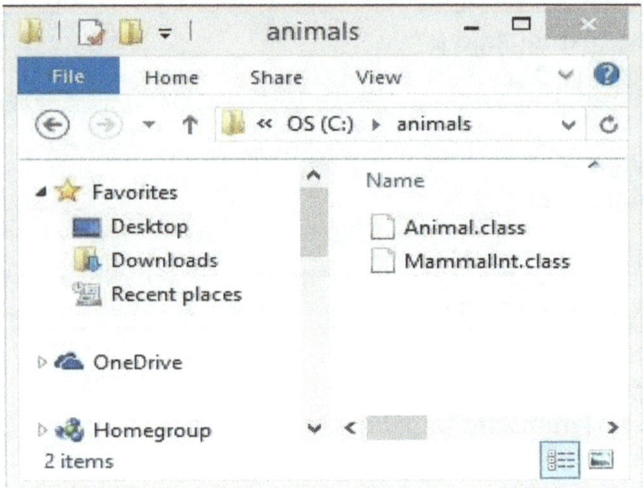

You can execute the class file within the package and get the result as shown below.

```
$ java animals.MammalInt
ammal eats
ammal travels
```

20.2 The Directory Structure of Packages

Two major results occur when a class is placed in a package:

- The name of the package becomes a part of the name of the class, as we just discussed in the previous section.
- The name of the package must match the directory structure where the corresponding bytecode resides.

Here is simple way of managing your files in Java:

Put the source code for a class, interface, enumeration, or annotation type in a text file whose name is the simple name of the type and whose extension is **.java.** For example:

```
// File Name : Car.java
package vehicle;
public class Car {
// Class implementation.
}
```

Now, put the source file in a directory whose name reflects the name of the package to which the class belongs:

```
....\vehicle\Car.java
```

Now, the qualified class name and pathname would be as follows:

- **Class name ->** vehicle.Car
- **Path name ->** vehicle\Car.java (in windows)

CHAPTER-20

JAVA- DATA STRUCTURES

The data structures provided by the Java utility package are very powerful and perform a wide range of functions. These data structures consist of the following interface and classes:

- Enumeration
- BitSet
- Vector
- Stack
- Dictionary
- Hashtable
- Properties

All these classes are now legacy and Java-2 has introduced a new framework called Collections Framework.

21.1 The Enumeration

The Enumeration interface isn't itself a data structure, but it is very important within the context of other data structures. The Enumeration interface defines a means to retrieve successive elements from a data structure.

For example, Enumeration defines a method called **nextElement** that is used to get the next element in a data structure that contains multiple elements.

21.2 The Enumeration Interface

The Enumeration interface defines the methods by which you can enumerate (obtain one at a time) the elements in a collection of objects.

This legacy interface has been superceded by Iterator. Although not deprecated, Enumeration is considered obsolete for new code. However, it is used by several methods defined by the legacy classes such as Vector and Properties, is used by several other API classes, and is currently in widespread use in application code.

The methods declared by Enumeration are summarized in the following table:

Table: 21.1

S.No	Methods with Description
1	**boolean hasMoreElements( )** When implemented, it must return true while there are still more elements to extract, and false when all the elements have been enumerated.
2	**Object nextElement( )** This returns the next object in the enumeration as a generic Object reference.

Example

Following is an example showing usage of Enumeration

```
import java.util.Vector;
import java.util.Enumeration;
public class EnumerationTester {
public static void main(String args[]) {
Enumeration days;
Vector dayNames = new Vector();
dayNames.add("Sunday");
dayNames.add("Monday");
dayNames.add("Tuesday");
dayNames.add("Wednesday");
dayNames.add("Thursday");
dayNames.add("Friday");
dayNames.add("Saturday");
days = dayNames.elements();
while (days.hasMoreElements()){
System.out.println(days.nextElement());
}
} }
```

This will produce the following result:

```
Sunday
Monday
Tuesday
Wednesday
Thursday
Friday
Saturday
```

21.3 The BitSet

The BitSet class implements a group of bits or flags that can be set and cleared individually.

This class is very useful in cases where you need to keep up with a set of Boolean values; you just assign a bit to each value and set or clear it as appropriate.

The BitSet Class

The BitSet class creates a special type of array that holds bit values. The BitSet array can increase in size as needed. This makes it similar to a vector of bits.

The BitSet defines the following two constructors.

Table: 21.2

S.No	Constructor and Description
1	**BitSet( )** This constructor creates a default object
2	**BitSet(int size)** This constructor allows you to specify its initial size, i.e., the number of bits that it can hold. All bits are initialized to zero

BitSet implements the Cloneable interface and defines the methods listed in the following table:

Table: 21.3

S.No	Methods with Description
1	**void and(BitSet bitSet)** ANDs the contents of the invoking BitSet object with those specified by bitSet. The result is placed into the invoking object.
2	**void andNot(BitSet bitSet)** For each 1 bit in bitSet, the corresponding bit in the invoking BitSet is cleared.
3	**int cardinality( )** Returns the number of set bits in the invoking object.
4	**void clear( )** Zeros all bits.
5	**void clear(int index)** Zeros the bit specified by index.
6	**void clear(int startIndex, int endIndex)** Zeros the bits from startIndex to endIndex.
7	**Object clone( )** Duplicates the invoking BitSet object.
8	**boolean equals(Object bitSet)** Returns true if the invoking bit set is equivalent to the one passed in bitSet. Otherwise, the method returns false.
9	**void flip(int index)** Reverses the bit specified by the index.
10	**void flip(int startIndex, int endIndex)** Reverses the bits from startIndex to endIndex.
11	**boolean get(int index)** Returns the current state of the bit at the specified index.
12	**BitSet get(int startIndex, int endIndex)** Returns a BitSet that consists of the bits from startIndex to endIndex. The invoking object is not changed.
13	**int hashCode( )** Returns the hash code for the invoking object.

14	**boolean intersects(BitSet bitSet)** Returns true if at least one pair of corresponding bits within the invoking object and bitSet are 1.
15	**boolean isEmpty( )** Returns true if all bits in the invoking object are zero.

Example

The following program illustrates several of the methods supported by this data structure:

```
import java.util.BitSet;
public class BitSetDemo {
public static void main(String args[]) {
BitSet bits1 = new BitSet(16);
BitSet bits2 = new BitSet(16);
// set some bits
for(int i=0; i<16; i++) {
if((i%2) == 0) bits1.set(i);
if((i%5) != 0) bits2.set(i);
}
System.out.println("Initial pattern in bits1: ");
System.out.println(bits1);
System.out.println("\nInitial pattern in bits2: ");
System.out.println(bits2);
```

Conti....

```
// AND bits
bits2.and(bits1);
System.out.println("\nbits2 AND bits1: ");
System.out.println(bits2);
// OR bits
bits2.or(bits1);
System.out.println("\nbits2 OR bits1: ");
System.out.println(bits2);
// XOR bits
bits2.xor(bits1);
System.out.println("\nbits2 XOR bits1: ");
System.out.println(bits2);
}
}
```

This will produce the following result:

```
Initial pattern in bits1:
{0, 2, 4, 6, 8, 10, 12, 14}
Initial pattern in bits2:
{1, 2, 3, 4, 6, 7, 8, 9, 11, 12, 13, 14}
bits2 AND bits1:
{2, 4, 6, 8, 12, 14}
bits2 OR bits1:
{0, 2, 4, 6, 8, 10, 12, 14}
bits2 XOR bits1:
{}
```

21.4 The Vector

The Vector class is similar to a traditional Java array, except that it can grow as necessary to accommodate new elements. Like an array, elements of a Vector object can be accessed via an index into the vector.

The nice thing about using the Vector class is that you don't have to worry about setting it to a specific size upon creation; it shrinks and grows automatically when necessary.

The Vector Class

Vector implements a dynamic array. It is similar to ArrayList, but with two differences:

- Vector is synchronized.
- Vector contains many legacy methods that are not part of the collections framework.

Vector proves to be very useful if you don't know the size of the array in advance or you just need one that can change sizes over the lifetime of a program.

Following is the list of constructors provided by the vector class.

Table: 21.4

S.No	Constructor and Description
1	**Vector( )** This constructor creates a default vector, which has an initial size of 10
2	**Vector(int size)** This constructor accepts an argument that equals to the required size, and creates a vector whose initial capacity is specified by size
3	**Vector(int size, int incr)** This constructor creates a vector whose initial capacity is specified by size and whose increment is specified by incr. The increment specifies the number of elements to allocate each time that a vector is resized upward
4	**Vector(Collection c)** This constructor creates a vector that contains the elements of collection c

Apart from the methods inherited from its parent classes, Vector defines the following methods:

Table: 21.5

S.No	Methods with Description
1	**void add(int index, Object element)** Inserts the specified element at the specified position in this Vector.
2	**boolean add(Object o)** Appends the specified element to the end of this Vector.
3	**boolean addAll(Collection c)** Appends all of the elements in the specified Collection to the end of this Vector, in the order that they are returned by the specified Collection's Iterator.
4	**boolean addAll(int index, Collection c)** Inserts all of the elements in in the specified Collection into this Vector at the specified position.
5	**void addElement(Object obj)** Adds the specified component to the end of this vector, increasing its size by one.
6	**int capacity()** Returns the current capacity of this vector.
7	**void clear()** Removes all of the elements from this vector.
8	**Object clone()** Returns a clone of this vector.
9	**boolean contains(Object elem)** Tests if the specified object is a component in this vector.
10	**boolean containsAll(Collection c)** Returns true if this vector contains all of the elements in the specified Collection.

Example

The following program illustrates several of the methods supported by this collection:

```
import java.util.*;
public class VectorDemo {
public static void main(String args[]) {
// initial size is 3, increment is 2
Vector v = new Vector(3, 2);
System.out.println("Initial size: " + v.size());
System.out.println("Initial capacity: " +
v.capacity());
v.addElement(new Integer(1));
v.addElement(new Integer(2));
v.addElement(new Integer(3));
v.addElement(new Integer(4));
System.out.println("Capacity after four additions: " +
v.capacity());
v.addElement(new Double(5.45));
System.out.println("Current capacity: " +
v.capacity());
v.addElement(new Double(6.08));
v.addElement(new Integer(7));
System.out.println("Current capacity: " +
```

Conti....

```
v.capacity());
v.addElement(new Float(9.4));
v.addElement(new Integer(10));
System.out.println("Current capacity: " +
v.capacity());
v.addElement(new Integer(11));
v.addElement(new Integer(12));
System.out.println("First element: " +
(Integer)v.firstElement());
System.out.println("Last element: " +
(Integer)v.lastElement());
if(v.contains(new Integer(3)))
System.out.println("Vector contains 3.");
// enumerate the elements in the vector.
Enumeration vEnum = v.elements();
System.out.println("\nElements in vector:");
while(vEnum.hasMoreElements())
System.out.print(vEnum.nextElement() + " ");
System.out.println();
}
}
```

This will produce the following result:

```
Initial size: 0
Initial capacity: 3
Capacity after four additions: 5
Current capacity: 5
Current capacity: 7
Current capacity: 9
First element: 1
Last element: 12
Vector contains 3.
Elements in vector:
1 2 3 4 5.45 6.08 7 9.4 10 11 12
```

21.5 The Stack

The Stack class implements a last-in-first-out (LIFO) stack of elements.

You can think of a stack literally as a vertical stack of objects; when you add a new element, it gets stacked on top of the others.

When you pull an element off the stack, it comes off the top. In other words, the last element you added to the stack is the first one to come back off.

The Stack Class

Stack is a subclass of Vector that implements a standard last-in, first-out stack.

Stack only defines the default constructor, which creates an empty stack. Stack includes all the methods defined by Vector, and adds several of its own.

```
Stack( )
```

Apart from the methods inherited from its parent class Vector, Stack defines the following methods:

Table: 21.6

S.No	Methods with Description
1	**boolean empty()** Tests if this stack is empty. Returns true if the stack is empty, and returns false if the stack contains elements.
2	**Object peek( )** Returns the element on the top of the stack, but does not remove it.
3	**Object pop( )** Returns the element on the top of the stack, removing it in the process.
4	**Object push(Object element)** Pushes the element onto the stack. Element is also returned.
5	**int search(Object element)** Searches for element in the stack. If found, its offset from the top of the stack is returned. Otherwise, .1 is returned.

Example

The following program illustrates several of the methods supported by this collection:

```
import java.util.*;
public class StackDemo {
static void showpush(Stack st, int a) {
st.push(new Integer(a));
System.out.println("push(" + a + ")");
System.out.println("stack: " + st);
}
static void showpop(Stack st) {
System.out.print("pop -> ");
Integer a = (Integer) st.pop();
System.out.println(a);
System.out.println("stack: " + st);
}
public static void main(String args[]) {
Stack st = new Stack();
System.out.println("stack: " + st);
showpush(st, 42);
showpush(st, 66);
showpush(st, 99);
showpop(st);
showpop(st);
showpop(st);
try {
showpop(st);
} catch (EmptyStackException e) {
System.out.println("empty stack");
}
}
}
```

This will produce the following result:

```
stack: [ ]
push(42)
stack: [42]
push(66)
stack: [42, 66]
push(99)
stack: [42, 66, 99]
pop -> 99
stack: [42, 66]
pop -> 66
stack: [42]
pop -> 42
stack: [ ]
pop -> empty stack
```

21.6 The Dictionary

The Dictionary class is an abstract class that defines a data structure for mapping keys to values.

This is useful in cases where you want to be able to access data via a particular key rather than an integer index.

Since the Dictionary class is abstract, it provides only the framework for a key-mapped data structure rather than a specific implementation.

The Dictionary Class

Dictionary is an abstract class that represents a key/value storage repository and operates much like Map.

Given a key and value, you can store the value in a Dictionary object. Once the value is stored, you can retrieve it by using its key. Thus, like a map, a dictionary can be thought of as a list of key/value pairs.

The abstract methods defined by Dictionary are listed below:

Table: 21.7

S.No	Methods with Description
1	**Enumeration elements( )** Returns an enumeration of the values contained in the dictionary.
2	**Object get(Object key)** Returns the object that contains the value associated with the key. If the key is not in the dictionary, a null object is returned.
3	**boolean isEmpty( )** Returns true if the dictionary is empty, and returns false if it contains at least one key.

4	**Enumeration keys( )** Returns an enumeration of the keys contained in the dictionary.
5	**Object put(Object key, Object value)** Inserts a key and its value into the dictionary. Returns null if the key is not already in the dictionary; returns the previous value associated with the key if the key is already in the dictionary.
6	**Object remove(Object key)** Removes the key and its value. Returns the value associated with the key. If the key is not in the dictionary, a null is returned.
7	**int size( )** Returns the number of entries in the dictionary.

21.7 The Map Interface

The Map interface maps unique keys to values. A key is an object that you use to retrieve a value at a later date.

- Given a key and a value, you can store the value in a Map object. After the value is stored, you can retrieve it by using its key.
- Several methods throw a NoSuchElementException when no items exist in the invoking map.
- A ClassCastException is thrown when an object is incompatible with the elements in a map.
- A NullPointerException is thrown if an attempt is made to use a null object and null is not allowed in the map.
- An UnsupportedOperationException is thrown when an attempt is made to change an unmodifiable map.

Table: 21.8

S.No	Methods with Description
1	**void clear( )** Removes all key/value pairs from the invoking map.
2	**boolean containsKey(Object k)** Returns true if the invoking map contains **k** as a key. Otherwise, returns false.
3	**boolean containsValue(Object v)** Returns true if the map contains **v** as a value. Otherwise, returns false.
4	**Set entrySet( )** Returns a Set that contains the entries in the map. The set contains objects of type Map.Entry. This method provides a set-view of the invoking map.
5	**boolean equals(Object obj)** Returns true if obj is a Map and contains the same entries. Otherwise, returns false.

6	**Object get(Object k)** Returns the value associated with the key k.
7	**int hashCode( )** Returns the hash code for the invoking map.
8	**boolean isEmpty( )** Returns true if the invoking map is empty. Otherwise, returns false.
9	**Set keySet( )** Returns a Set that contains the keys in the invoking map. This method provides a set-view of the keys in the invoking map.
10	**void putAll(Map m)** Puts all the entries from **m** into this map.

Example

Map has its implementation in various classes like HashMap. Following is an example to explain map functionality:

```
import java.util.*;
public class CollectionsDemo {
public static void main(String[] args) {
Map m1 = new HashMap();
m1.put("Zara", "8");
m1.put("Mahnaz", "31");
m1.put("Ayan", "12");
m1.put("Daisy", "14");
System.out.println();
System.out.println(" Map Elements");
System.out.print("\t" + m1);
}
}
```

21.8 The Hashtable

The Hashtable class provides a means of organizing data based on some user-defined key structure.

For example, in an address list hash table you could store and sort data based on a key such as ZIP code rather than on a person's name.

The specific meaning of keys with regard to hash tables is totally dependent on the usage of the hash table and the data it contains.

The Hashtable Class

Hashtable was part of the original java.util and is a concrete implementation of a Dictionary.

However, Java 2 re-engineered Hashtable so that it also implements the Map interface. Thus, Hashtable is now integrated into the collections framework. It is similar to HashMap, but is synchronized.

Like HashMap, Hashtable stores key/value pairs in a hash table. When using a Hashtable, you specify an object that is used as a key, and the value that you want linked to that key. The key is then hashed, and the resulting hash code is used as the index at which the value is stored within the table.

Following is the list of constructors provided by the HashTable class.

Table: 21.9

S.No	Constructor and Description
1	**Hashtable( )** This is the default constructor of the hash table it instantiates the Hashtable class.
2	**Hashtable(int size)** This constructor accepts an integer parameter and creates a hash table that has an initial size specified by integer value size.
3	**Hashtable(int size, float fillRatio)** This creates a hash table that has an initial size specified by size and a fill ratio specified by fillRatio. This ratio must be between 0.0 and 1.0, and it determines how full the hash table can be before it is resized upward.
4	**Hashtable(Map<? extends K, ? extends V> t)** This constructs a Hashtable with the given mappings.

Example

The following program illustrates several of the methods supported by this data structure:

```
import java.util.*;
public class HashTableDemo {
public static void main(String args[]) {
// Create a hash map
Hashtable balance = new Hashtable();
Enumeration names;
String str;
double bal;
balance.put("Zara", new Double(3434.34));
balance.put("Mahnaz", new Double(123.22));
balance.put("Ayan", new Double(1378.00));
balance.put("Daisy", new Double(99.22));
balance.put("Qadir", new Double(-19.08));
// Show all balances in hash table.
names = balance.keys();
while(names.hasMoreElements()) {
str = (String) names.nextElement();
System.out.println(str + ": " + balance.get(str));
}
System.out.println();
// Deposit 1,000 into Zara's account
bal = ((Double)balance.get("Zara")).doubleValue();
balance.put("Zara", new Double(bal+1000));
System.out.println("Zara's new balance: " +
balance.get("Zara"));
}
}
```

This will produce the following result:

```
Qadir: -19.08

Zara: 3434.34

ahnaz: 123.22

Daisy: 99.22

Ayan: 1378.0

Zara's new balance: 4434.34
```

21.9 The Properties

A property is a subclass of Hashtable. It is used to maintain lists of values in which the key is a String and the value is also a String.

The Properties class is used by many other Java classes. For example, it is the type of object returned by **System.getProperties( )** when obtaining environmental values.

The Properties Class

Properties define the following instance variable. This variable holds a default property list associated with a Properties object.

```
Properties defaults;
```

Following is the list of constructors provided by the properties class.

Table: 21.10

S.No	Constructors and Description
1	**Properties( )** This constructor creates a Properties object that has no default values.
2	**Properties(Properties propDefault)** Creates an object that uses propDefault for its default values. In both cases, the property list is empty.

CHAPTER - 21

JAVA UTILITIES

Java.util package contains the collections framework, legacy collection classes, event model, date and time facilities, internationalization, and miscellaneous utility classes. This reference will take you through simple and practical methods available in java.util package.

22.1 Collections

Java collections refer to a single unit of objects. You can perform all operations on data such as searching, sorting, insertion, manipulation, deletion, etc. by Java collections.

The java.util.Collections class consists exclusively of static methods that operate on or return collections. Following are the important points about Collections:

- It contains polymorphic algorithms that operate on collections, "wrappers", which return a new collection backed by a specified collection.
- The methods of this class all throw a NullPointerException if the collections or class objects provided to them are null.

The collections framework was designed to meet several goals, such as:

- The framework had to be high-performance. The implementations for the fundamental collections (dynamic arrays, linked lists, trees, and hashtables) were to be highly efficient.
- The framework had to allow different types of collections to work in a similar manner and with a high degree of interoperability.
- The framework had to extend and/or adapt a collection easily.

Class Declaration

Following is the declaration for java.util.Collections class:

```
public class Collections
  extends Object
```

Field:

Following are the fields for java.util.Collections class:

- Static List **EMPTY_LIST** -- This is the empty list (immutable).
- Static Map **EMPTY_MAP** -- This is the empty map (immutable).
- Static Set **EMPTY_SET** -- This is the empty set (immutable).

22.2 Java Collection Framework

A Java collection framework provides architecture to store and manipulate a group of objects. A Java collection framework includes the following:

- **Interfaces**
- **Classes**
- **Algorithm**

Interfaces:

- Interface in Java refers to the abstract data types.
- They allow Java collections to be manipulated independently from the details of their representation.
- They form a hierarchy in object-oriented programming languages.

Classes:

- Classes in Java are the implementation of the collection interface.
- It basically refers to the data structures that are used again and again.

Algorithm:

- Algorithm refers to the methods which are used to perform operations such as searching and sorting, on objects that implement collection interfaces.
- Algorithms are polymorphic in nature as the same method can be used to take many forms or you can say perform different implementations of the Java collection interface.

The Java collection framework provides the developers to access pre-packaged data structures as well as algorithms to manipulate data. Next, let us move to the Java collections framework hierarchy and see where these interfaces and classes reside.

22.3 Java Collection Framework Hierarchy

As we have learned Java collection framework includes interfaces and classes. Now, let us see the Java collections framework hierarchy.

Fig: 22.1

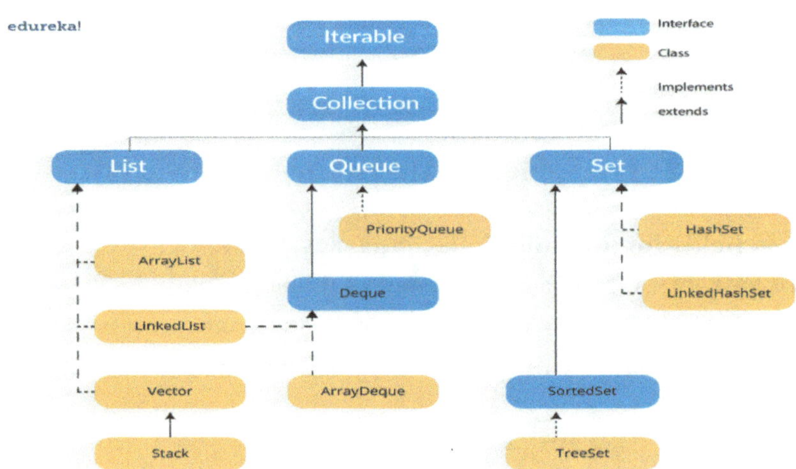

In the above image, blue part refers to the different interfaces and the yellow part defines the class. Now, let us understand these components in detail.

1) Java Collections: Interface

Iterator interface: Iterator is an interface that iterates the elements. It is used to traverse the list and modify the elements. Iterator interface has three methods which are mentioned below:

- public boolean hasNext() – This method returns true if iterator has more elements.
- public object next() – It returns the element and moves the cursor pointer to the next element.
- public void remove() – This method removes the last elements returned by the iterator.

There are three components that extend the collection interface i.e List, Queue and Sets. Let's learn about them in detail:

2) Java Collections: List

A List is an ordered Collection of elements which may contain duplicates. It is an interface that extents the Collection interface. Lists are further classified into the following:

- **java.util.ArrayList**
- **java.util.LinkedList**
- **java.util.Vector**

i) java.util.Array list:

ArrayList is the implementation of List Interface where the elements can be dynamically added or removed from the list. Also, the size of the list is increased dynamically if the elements are added more than the initial size.

Fig: 22.2

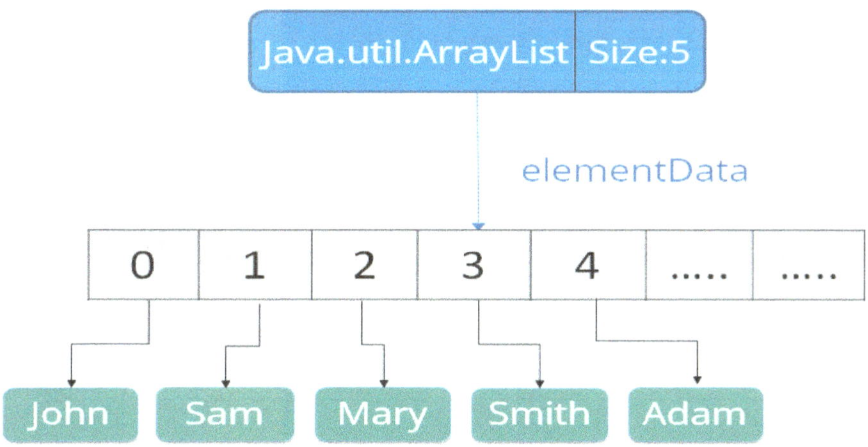

Syntax:

```
ArrayList object = new ArrayList ();
```

Some of the methods in array list are listed below:

Table: 22.1

Method	Description
boolean add(Collection c)	Appends the specified element to the end of a list.
void add(int index, Object element)	Inserts the specified element at the specified position.
void clear()	Removes all the elements from this list.
int lastIndexOf(Object o)	Return the index in this list of the last occurrence of the specified element, or -1 if the list does not contain this element.
Object clone()	Return a shallow copy of an ArrayList.
Object[] toArray()	Returns an array containing all the elements in the list.
void trimToSize()	Trims the capacity of this ArrayList instance to be the list's current size.

Array list with a programmatic example:

```
import java.util.*;
 class ArrayListExample{
 public static void main(String args[]){

 ArrayList al=new ArrayList();  // creating array list
 al.add("Jack");                // adding elements
 al.add("Tyler");
 Iterator itr=al.iterator();
 while(itr.hasNext()){
 System.out.println(itr.next());
 }
 }
 }
```

ii) java.util.Linked List:

- Linked List is a sequence of links which contains items. Each link contains a connection to another link.

Syntax: Linkedlist object = new Linkedlist();

Java Linked List class uses two types of Linked list to store the elements:

- Singly Linked List
- Doubly Linked List

Singly Linked List:

- In a singly linked list each node in this list stores the data of the node and a pointer or reference to the next node in the list. Refer to the below image to get a better understanding of single Linked list.

Fig: 22.3

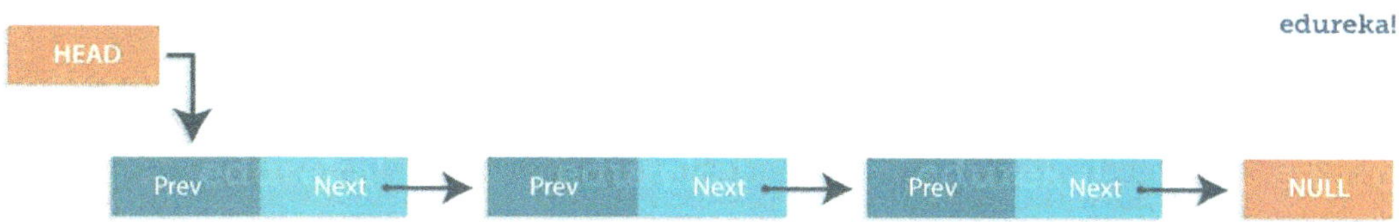

Doubly Linked List:

- In a doubly Linked list, it has two references, one to the next node and another to previous node. You can refer to the below image to get a better understanding of doubly linked list.

Fig: 22.4

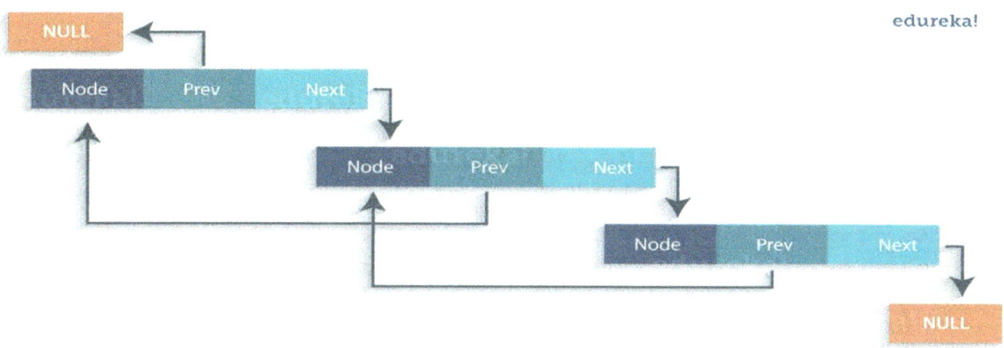

Some of the methods in the linked list are listed below:

Table: 22.2

Method	Description
boolean add(Object o)	It is used to append the specified element to the end of the vector.
boolean contains(Object o)	Returns true if this list contains the specified element.
void add (int index, Object element)	Inserts the element at the specified element in the vector.
void addFirst(Object o)	It is used to insert the given element at the beginning.
void addLast(Object o)	It is used to append the given element to the end.
int size()	It is used to return the number of elements in a list

boolean remove(Object o)	Removes the first occurrence of the specified element from this list.
int indexOf(Object element)	Returns the index of the first occurrence of the specified element in this list, or -1.
int lastIndexOf(Object element)	Returns the index of the last occurrence of the specified element in this list, or -1.

Linked list with a programmatic example:

```
import java.util.*;
public class LinkedlistExample{
public static void main(String args[]){
LinkedList<String> al=new LinkedList<String>(); // creating linked list
al.add("Rachit");                                // adding elements
al.add("Rahul");
al.add("Rajat");
Iterator<String> itr = al.iterator();
while(itr.hasNext()){
System.out.println(itr.next());
 }
 }
 }
```

iii) java.util.Vector:

- Vectors are similar to arrays, where the elements of the vector object can be accessed via an index into the vector.
- Vector implements a dynamic array. Also, the vector is not limited to a specific size, it can shrink or grow automatically whenever required.

It is similar to ArrayList, but with two differences:

- Vector is synchronized.
- Vector contains many legacy methods that are not part of the collections framework.

Fig: 22.5

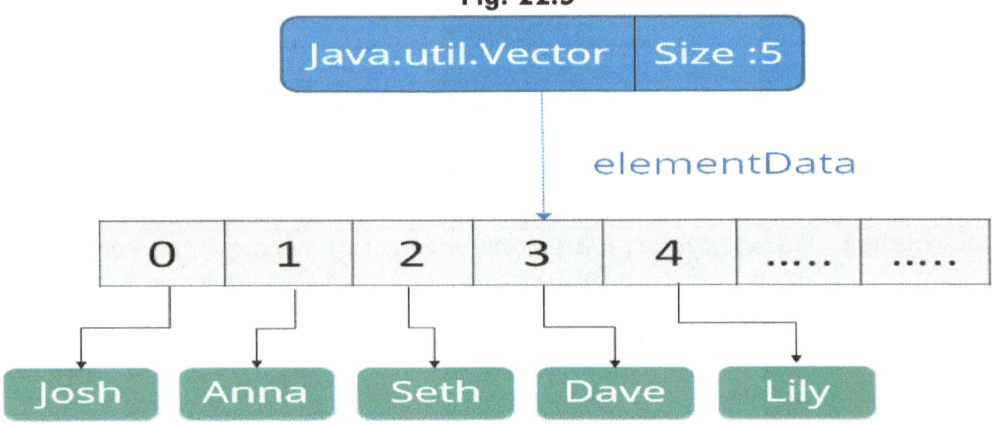

Syntax:

```
Vector object = new Vector(size,increment);
```

Below are some of the methods of the Vector class:

Table: 22.3

Method	Description
boolean add(Object o)	Appends the specified element to the end of the list.
void clear()	Removes all of the elements from this list.
void add(int index, Object element)	Inserts the specified element at the specified position.
boolean remove(Object o)	Removes the first occurrence of the specified element from this list.
boolean contains(Object element)	Returns true if this list contains the specified element.
int indexOfObject (Object element)	Returns the index of the first occurrence of the specified element in the list, or -1.
int size()	Returns the number of elements in this list.
int lastIndexOf(Object o)	int lastIndexOf(Object o)

Java collections example:

```
package collections;

import java.util.ArrayList;
import java.util.Arrays;
import java.util.List;

public class MyArrayList {

    public static void main(String[] args) {

        // create a list using the
        List<String> list = Arrays.asList("Lars", "Simon");
        // alternatively
        List<String> anotherList = new ArrayList<>();
        anotherList.add("Lars");
        anotherList.add("Simon");
        // print each element to the console using method references
        list.forEach(System.out::println);
        anotherList.forEach(System.out::println);

    }
```

22.4 I/O Streams

- The java.io package contains nearly every class you might ever need to perform input and output (I/O) in Java.
- All these streams represent an input source and an output destination.
- The stream in the java.io package supports many data such as primitives, object, localized characters, etc.

i) What Is Stream?

A stream can be defined as a sequence of data. There are two kinds of Streams:

- **InPutStream:** The InputStream is used to read data from a source.
- **OutPutStream:** The OutputStream is used for writing data to a destination.

Fig: 22.6

Java provides strong but flexible support for I/O related to files and networks but this tutorial covers very basic functionality related to streams and I/O. We will see the most commonly used examples one by one:

ii) What Is Byte Streams?

- Java byte streams are used to perform input and output of 8-bit bytes.
- Though there are many classes related to byte streams but the most frequently used classes are, **FileInputStream and FileOutputStream.**

Following is an example which makes use of these two classes to copy an input file into an output file:

```
import java.io.*;
public class CopyFile {
public static void main(String args[]) throws IOException
 {
        FileInputStream in = null;
        FileOutputStream out = null;
try {
        in = new FileInputStream("input.txt");
out = new FileOutputStream("output.txt");
int c;
while ((c = in.read()) != -1) {
        out.write(c);
 }
}finally {
        if (in != null) {
                in.close();
}
if (out != null) {
         out.close();
        }
      }
   }
}
```

Now let's have a file **input.txt** with the following content:

```
This is test for copy file.
```

As a next step, compile the above program and execute it, which will result in creating output.txt file with the same content as we have in input.txt. So let's put the above code in **CopyFile.java file** and do the following:

```
$javac CopyFile.java
$java CopyFile
```

iii) Character Streams

- Java Byte streams are used to perform input and output of 8-bit bytes, whereas Java Character streams are used to perform input and output for 16-bit unicode.
- Though there are many classes related to character streams but the most frequently used classes are, FileReader and FileWriter.

- Though internally FileReader uses FileInputStream and FileWriter uses FileOutputStream but here the major difference is that FileReader reads two bytes at a time and FileWriter writes two bytes at a time.

iv) Standard Stream

All the programming languages provide support for standard I/O where the user's program can take input from a keyboard and then produce an output on the computer screen. If you are aware of C or C++ programming languages, then you must be aware of three standard devices STDIN, STDOUT and STDERR. Similarly, Java provides the following three standard streams:

- **Standard Input:** This is used to feed the data to user's program and usually a keyboard is used as standard input stream and represented as System.in.
- **Standard Output:** This is used to output the data produced by the user's program and usually a computer screen is used for standard output stream and represented as System.out.
- **Standard Error:** This is used to output the error data produced by the user's program and usually a computer screen is used for standard error stream and represented as System.err.

Following is a simple program, which creates InputStreamReader to read standard input stream until the user types a "q":

```
import java.io.*;
public class ReadConsole {
   public static void main(String args[]) throws IOException
      {
         InputStreamReader cin = null;
         try {
                  cin = new InputStreamReader(System.in);
                  System.out.println("Enter characters, 'q' to quit.");
                  char c;
                  do {
                      c = (char) cin.read();
                    System.out.print(c);
                   } while(c != 'q');
         }finally {
                   if (cin != null) {
                       cin.close();
                     }
               }
        }
}
```

Let's keep the above code in ReadConsole.java file and try to compile and execute it as shown in the following program. This program continues to raed and output the same character until we press 'q':

```
$javac ReadConsole.java
$java ReadConsole
Enter characters, 'q' to quit.
1
1
e
e
q
q
```

22.5 Reading and Writing Files

As described earlier, a stream can be defined as a sequence of data. The **InputStream** is used to read data from a source and the **OutputStream** is used for writing data to a destination. Here is a hierarchy of classes to deal with Input and Output streams.

Fig: 22.7

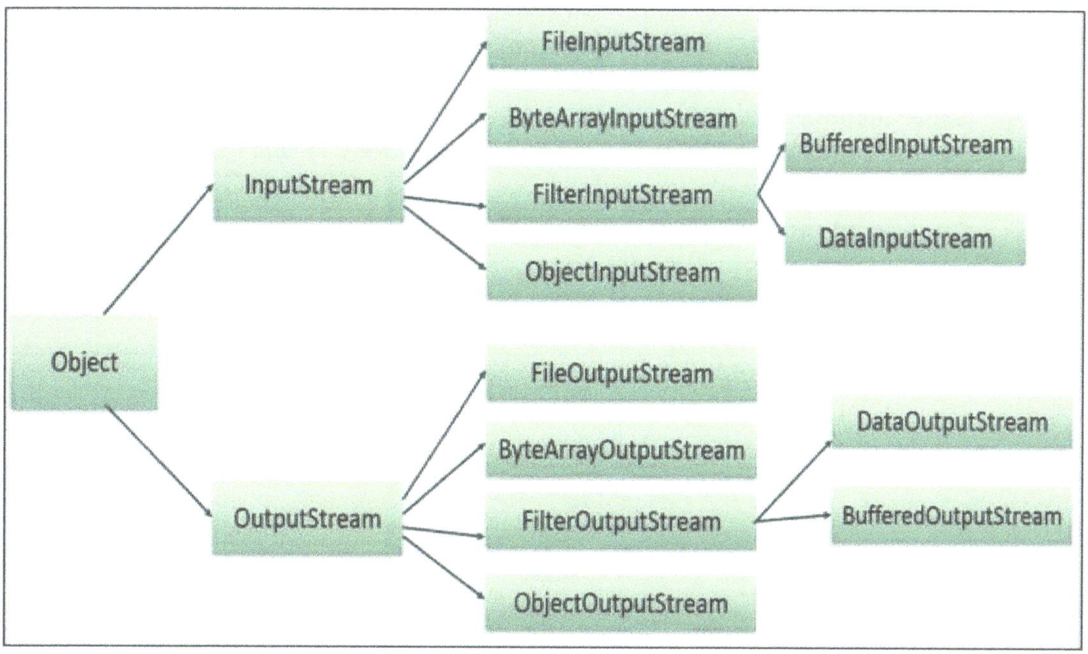

The two important streams are **FileInputStream** and **FileOutputStream,** which would be discussed in this chapter.

1) Fileinputstream

This stream is used for reading data from the files. Objects can be created using the keyword new and there are several types of constructors available.

Following constructor takes a file name as a string to create an input stream object to read the file:

```
InputStream f = new FileInputStream("C:/java/hello");
```

Following constructor takes a file object to create an input stream object to read the file. First we create a file object using File() method as follows:

```
File f = new File("C:/java/hello");
InputStream f = new FileInputStream(f);
```

Once you have InputStream object in hand, then there is a list of helper methods which can be used to read to stream or to do other operations on the stream.

Table: 22.4

S.No	Methods with Description
1	**public void close() throws IOException{}** This method closes the file output stream. Releases any system resources associated with the file. Throws an IOException.
2	**protected void finalize()throws IOException {}** This method cleans up the connection to the file. Ensures that the close method of this file output stream is called when there are no more references to this stream. Throws an IOException.
3	**public int read(int r)throws IOException{}** This method reads the specified byte of data from the InputStream. Returns an int. Returns the next byte of data and -1 will be returned if it's the end of the file.
4	**public int read(byte[] r) throws IOException{}** This method reads r.length bytes from the input stream into an array. Returns the total number of bytes read. If it is the end of the file, -1 will be returned.
5	**public int available() throws IOException{}** Gives the number of bytes that can be read from this file input stream. Returns an int.

There are other important input streams available, for more detail you can refer to the following links:

- **ByteArrayInputStream**
- **DataInputStream**

i) Bytearrayinputstream

- The ByteArrayInputStream class allows a buffer in the memory to be used as an InputStream.
- The input source is a byte array.

ByteArrayInputStream class provides the following constructors.

Table: 22.5

S.No	Constructor and Description
1	**ByteArrayInputStream(byte [] a)** This constructor accepts a byte array as a parameter.
2	**ByteArrayInputStream(byte [] a, int off, int len)** This constructor takes an array of bytes, and two integer values, where **off** is the first byte to be read and **len** is the number of bytes to be read.

Once you have **ByteArrayInputStream** object in hand then there is a list of helper methods which can be used to read the stream or to do other operations on the stream.

Table: 22.6

S.No	Methods with Description
1	**public int read()** This method reads the next byte of data from the InputStream. Returns an int as the next byte of data. If it is the end of the file, then it returns -1.
2	**public int read(byte[] r, int off, int len)** This method reads upto **len** number of bytes starting from **off** from the input stream into an array. Returns the total number of bytes read. If it is the end of the file, -1 will be returned.
3	**public int available()** Gives the number of bytes that can be read from this file input stream. Returns an int that gives the number of bytes to be read.
4	**public void mark(int read)** This sets the current marked position in the stream. The parameter gives the maximum limit of bytes that can be read before the marked position becomes invalid.
5	**public long skip(long n)** Skips 'n' number of bytes from the stream. This returns the actual number of bytes skipped.

Example

Following is the example to demonstrate ByteArrayInputStream and ByteArrayOutputStream.

```
import java.io.*;
public class ByteStreamTest {
 public static void main(String args[])throws IOException {
     ByteArrayOutputStream bOutput = new ByteArrayOutputStream(12);
            while( bOutput.size()!= 10 ) {
               // Gets the inputs from the user
               bOutput.write(System.in.read());
            }
            byte b [] = bOutput.toByteArray();
            System.out.println("Print the content");
            for(int x= 0 ; x < b.length; x++) {
               // printing the characters

            System.out.print((char)b[x] + " ");
 }
 System.out.println(" ");
 int c;
 ByteArrayInputStream bInput = new ByteArrayInputStream(b);
 System.out.println("Converting characters to Upper case " );
 for(int y = 0 ; y < 1; y++ ) {
            while(( c= bInput.read())!= -1) {
                System.out.println(Character.toUpperCase((char)c));
            }
                bInput.reset();
   }
  }
}
```

Following is the sample run of the above program:

```
Asdfghjkly
Print the content
a s d f g h j k l y
Converting characters to Upper case
A
S
D
F
G
H
J
K
L
Y
```

ii) Datainputstream

The DataInputStream is used in the context of DataOutputStream and can be used to read primitives.

Following is the constructor to create an InputStream:

InputStream in = DataInputStream (InputStream in);

Once you have DataInputStream object in hand, then there is a list of helper methods, which can be used to read the stream or to do other operations on the stream.

Table: 22.7

S.No	Methods with Description
1	**public final int read(byte[] r, int off, int len)throws IOException** Reads up to len bytes of data from the input stream into an array of bytes. Returns the total number of bytes read into the buffer otherwise -1 if it is end of file.
2	**Public final int read(byte [] b)throws IOException** Reads some bytes from the inputstream an stores in to the byte array. Returns the total number of bytes read into the buffer otherwise -1 if it is end of file.
3	**(a) public final Boolean readBooolean()throws IOException** **(b) public final byte readByte()throws IOException** **(c) public final short readShort()throws IOException** **(d) public final Int readInt()throws IOException** These methods will read the bytes from the contained InputStream. Returns the next two bytes of the InputStream as the specific primitive type.
4	**public String readLine() throws IOException** Reads the next line of text from the input stream. It reads successive bytes, converting each byte separately into a character, until it encounters a lineterminator or end of file; the characters read are then returned as a String.

Example:

Following is an example to demonstrate DataInputStream and DataOutputStream. This example reads 5 lines given in a file test.txt and converts those lines into capital letters and finally copies them into another file test1.txt.

```
import java.io.*;
public class DataInput_Stream{
    public static void main(String args[])throws IOException{

        //writing string to a file encoded as modified UTF-8
                DataOutputStream dataOut = new DataOutputStream(new
         FileOutputStream("E:\\file.txt"));
                        dataOut.writeUTF("hello");

        //Reading data from the same file
                        DataInputStream dataIn = new DataInputStream(new
         FileInputStream("E:\\file.txt"));

                        while(dataIn.available()>0){

                            String k = dataIn.readUTF();
                            System.out.print(k+" ");
                        }

                }
        }
```

Following is the sample run of the above program:

```
Hello
```

2) Fileoutputstream

- FileOutputStream is used to create a file and write data into it. The stream would create a file, if it doesn't already exist, before opening it for output.
- Here are two constructors which can be used to create a FileOutputStream object.

Following constructor takes a file name as a string to create an input stream object to write the file:

OutputStream f = new FileOutputStream("C:/java/hello")

Following constructor takes a file object to create an output stream object to write the file. First, we create a file object using File() method as follows:

```
File f = new File("C:/java/hello");
OutputStream f = new FileOutputStream(f);
```

Once you have OutputStream object in hand, then there is a list of helper methods, which can be used to write to stream or to do other operations on the stream.

Table: 22.8

S.No	Methods with Description
1	**public void close() throws IOException{}** This method closes the file output stream. Releases any system resources associated with the file. Throws an IOException.
2	**protected void finalize()throws IOException {}** This method cleans up the connection to the file. Ensures that the close method of this file output stream is called when there are no more references to this stream. Throws an IOException.
3	**public void write(int w)throws IOException{}** This method writes the specified byte to the output stream.
4	**public void write(byte[] w)** Writes w.length bytes from the mentioned byte array to the OutputStream.

There are other important output streams available, for more detail you can refer to the following links:

- **ByteArrayOutputStream**
- **DataOutputStream**

i) ByteArrayOutputStream

The **ByteArrayOutputStream** class stream creates a buffer in memory and all the data sent to the stream is stored in the buffer.

Following is the list of the constructors to be provided by **ByteArrayOutputStream** class.

S.No	Constructors and Description
1	**ByteArrayOutputStream()** This constructor creates a ByteArrayOutputStream having buffer of 32 byte
2	**ByteArrayOutputStream(int a)** This constructor creates a ByteArrayOutputStream having buffer of the given size

Once you have ByteArrayOutputStream object in hand, then there is a list of helper methods which can be used to write the stream or to do other operations on the stream.

Table: 22.9

S.No	Methods with Description
1	**public void reset()** This method resets the number of valid bytes of the byte array output stream to zero, so all the accumulated output in the stream will be discarded.
2	**public byte[] toByteArray()** This method creates a newly allocated Byte array. Its size would be the current size of the output stream and the contents of the buffer will be copied into it. Returns the current contents of the output stream as a byte array.
3	**public String toString()** Converts the buffer content into a string. Translation will be done according to the default character encoding. Returns the String translated from the buffer's content.
4	**public void write(int w)** Writes the specified array to the output stream.
5	**public void write(byte []b, int of, int len)** Writes len number of bytes starting from offset off to the stream.
6	**public void writeTo(OutputStream outSt)** Writes the entire content of this Stream to the specified stream argument.

Example

Following is an example to demonstrate ByteArrayOutputStream and ByteArrayInputStream.

```
import java.io.*;
public class ByteStreamTest {
   public static void main(String args[])throws IOException {
      ByteArrayOutputStream bOutput = new ByteArrayOutputStream(12);
         while( bOutput.size()!= 10 ) {
            // Gets the inputs from the user
               bOutput.write(System.in.read());
         }
            byte b [] = bOutput.toByteArray();
             System.out.println("Print the content");
             for(int x= 0 ; x < b.length; x++) {
            //printing the characters
              System.out.print((char)b[x] + " ");
             }
               System.out.println(" ");
             int c;
             ByteArrayInputStream bInput = new ByteArrayInputStream(b);
              System.out.println("Converting characters to Upper case " );
             for(int y = 0 ; y < 1; y++ ) {
                while(( c= bInput.read())!= -1) {
                System.out.println(Character.toUpperCase((char)c));
              }
            bInput.reset();
           }
      }
}
```

Here is the sample run of the above program:

```
Asdfghjkly
Print the content
a s d f g h j k l y
Converting characters to Upper case
      A
      S
      D
      F
      G
      H
      J
      K
      L
      Y
```

ii) Dataoutputstream

The DataOutputStream stream lets you write the primitives to an output source.

Following is the constructor to create a DataOutputStream.

DataOutputStream out = DataOutputStream(OutputStream out);

Once you have DataOutputStream object in hand, then there is a list of helper methods, which can be used to write the stream or to do other operations on the stream.

Table: 22.10

S.No	Methods with Description
1	**public final void write(byte[] w, int off, int len)throws IOException** Writes len bytes from the specified byte array starting at point off, to the underlying stream.
2	**Public final int write(byte [] b)throws IOException** Writes the current number of bytes written to this data output stream. Returns the total number of bytes written into the buffer.
3	**(a) public final void writeBooolean()throws IOException,** **(b) public final void writeByte()throws IOException,** **(c) public final void writeShort()throws IOException** **(d) public final void writeInt()throws IOException** These methods will write the specific primitive type data into the output stream as bytes.
4	**Public void flush()throws IOException** Flushes the data output stream.
5	**public final void writeBytes(String s) throws IOException** Writes out the string to the underlying output stream as a sequence of bytes. Each character in the string is written out, in sequence, by discarding its high eight bits.

Example:1

Following is an example to demonstrate DataInputStream and DataOutputStream. This example reads 5 lines given in a file test.txt and converts those lines into capital letters and finally copies them into another file test1.txt.

```
import java.io.*;
public class DataInput_Stream{
    public static void main(String args[])throws IOException{

  //writing string to a file encoded as modified UTF-8
 DataOutputStream dataOut = new DataOutputStream(new
    FileOutputStream("E:\\file.txt"));
    dataOut.writeUTF("hello");

  //Reading data from the same file
   DataInputStream dataIn = new DataInputStream(new
      FileInputStream("E:\\file.txt"));

            while(dataIn.available()>0){

                    String k = dataIn.readUTF();
                    System.out.print(k+" ");
            }

      }
  }
```

Here is the sample run of the above program:

```
THIS IS TEST 1 ,
THIS IS TEST 2 ,
THIS IS TEST 3 ,
THIS IS TEST 4 ,
THIS IS TEST 5 ,
```

Example: 2

Following is the example to demonstrate InputStream and OutputStream:

```
import java.io.*;
public class fileStreamTest{
    public static void main(String args[]){

        try{
                byte bWrite [] = {11,21,3,40,5};
                OutputStream os = new FileOutputStream("test.txt");
                for(int x=0; x < bWrite.length ; x++){
                    os.write( bWrite[x] ); // writes the bytes
            }
            os.close();

            InputStream is = new FileInputStream("test.txt");
            int size = is.available();
            for(int i=0; i< size; i++){
                    System.out.print((char)is.read() + " ");
            }
            is.close();
 }catch(IOException e){
            System.out.print("Exception");
    }
  }
}
```

The above code would create file test.txt and would write given numbers in binary format. Same would be the output on the stdout screen.

22.6 File Navigation and I/O

There are several other classes that we would be going through to get to know the basics of File Navigation and I/O.

- File Class
- FileReader Class
- FileWriter Class

i) File Class

- Java File class represents the files and directory pathnames in an abstract manner. This class is used for creation of files and directories, file searching, file deletion, etc.
- The File object represents the actual file/directory on the disk. Following is the list of constructors to create a File object.

Fig: 22.11

S.No	Methods with Description
1	**File(File parent, String child)** This constructor creates a new File instance from a parent abstract pathname and a child pathname string.
2	**File(String pathname)** This constructor creates a new File instance by converting the given pathname string into an abstract pathname.
3	**File(String parent, String child)** This constructor creates a new File instance from a parent pathname string and a child pathname string.
4	**File(URI uri)** This constructor creates a new File instance by converting the given file: URI into an abstract pathname.

Once you have File object in hand, then there is a list of helper methods which can be used to manipulate the files.

Fig: 22.12

S.No	Methods with Description
1	**public String getName()** Returns the name of the file or directory denoted by this abstract pathname.
2	**public String getParent()** Returns the pathname string of this abstract pathname's parent, or null if this pathname does not name a parent directory.
3	**public File getParentFile()** Returns the abstract pathname of this abstract pathname's parent, or null if this pathname does not name a parent directory.
4	**public String getPath()** Converts this abstract pathname into a pathname string.
5	**public boolean isAbsolute()** Tests whether this abstract pathname is absolute. Returns true if this abstract pathname is absolute, false otherwise.
6	**public String getAbsolutePath()** Returns the absolute pathname string of this abstract pathname.
7	**public boolean canRead()** Tests whether the application can read the file denoted by this abstract pathname. Returns true if and only if the file specified by this abstract pathname exists and can be read by the application; false otherwise.
8	**public boolean canRead()** Tests whether the application can read the file denoted by this abstract

	pathname. Returns true if and only if the file specified by this abstract pathname exists and can be read by the application; false otherwise.
9	**public boolean exists()** Tests whether the file or directory denoted by this abstract pathname exists. Returns true if and only if the file or directory denoted by this abstract pathname exists; false otherwise.
10	**public boolean isDirectory()** Tests whether the file denoted by this abstract pathname is a directory. Returns true if and only if the file denoted by this abstract pathname exists and is a directory; false otherwise.
11	**public boolean isFile()** Tests whether the file denoted by this abstract pathname is a normal file. A file is normal if it is not a directory and, in addition, satisfies other systemdependent criteria. Any non-directory file created by a Java application is guaranteed to be a normal file. Returns true if and only if the file denoted by this abstract pathname exists and is a normal file; false otherwise.
12	**public long lastModified()** Returns the time that the file denoted by this abstract pathname was last modified. Returns a long value representing the time the file was last modified, measured in milliseconds since the epoch (00:00:00 GMT, January 1, 1970), or 0L if the file does not exist or if an I/O error occurs.
13	**public long length()** Returns the length of the file denoted by this abstract pathname. The return value is unspecified if this pathname denotes a directory.
14	**public boolean createNewFile() throws IOException** Atomically creates a new, empty file named by this abstract pathname if and only if a file with this name does not yet exist. Returns true if the named file does not exist and was successfully created; false if the named file already exists.
15	**public boolean delete()** Deletes the file or directory denoted by this abstract pathname. If this pathname denotes a directory, then the directory must be empty in order to be deleted. Returns true if and only if the file or directory is successfully deleted; false otherwise.
16	**public void deleteOnExit()** Requests that the file or directory denoted by this abstract pathname be deleted when the virtual machine terminates.
17	**public String[] list()** Returns an array of strings naming the files and directories in the directory denoted by this abstract pathname.

18	**public String[] list(FilenameFilter filter)** Returns an array of strings naming the files and directories in the directory denoted by this abstract pathname that satisfy the specified filter.
19	**public File[] listFiles()** Returns an array of abstract pathnames denoting the files in the directory denoted by this abstract pathname.
20	**public File[] listFiles(FileFilter filter)** Returns an array of abstract pathnames denoting the files and directories in the directory denoted by this abstract pathname that satisfy the specified filter.
21	**public boolean mkdir()** Creates the directory named by this abstract pathname. Returns true if and only if the directory was created; false otherwise.
22	**public boolean mkdirs()** Creates the directory named by this abstract pathname, including any necessary but nonexistent parent directories. Returns true if and only if the directory was created, along with all necessary parent directories; false otherwise.
23	**public boolean renameTo(File dest)** Renames the file denoted by this abstract pathname. Returns true if and only if the renaming succeeded; false otherwise.
24	**public boolean setLastModified(long time)** Sets the last-modified time of the file or directory named by this abstract pathname. Returns true if and only if the operation succeeded; false otherwise.
25	**public boolean setReadOnly()** Marks the file or directory named by this abstract pathname so that only read operations are allowed. Returns true if and only if the operation succeeded; false otherwise.
26	**public static File createTempFile(String prefix, String suffix, File directory) throws IOException** Creates a new empty file in the specified directory, using the given prefix and suffix strings to generate its name. Returns an abstract pathname denoting a newly-created empty file.
27	**public static File createTempFile(String prefix, String suffix) throws IOException** Creates an empty file in the default temporary-file directory, using the given prefix and suffix to generate its name. Invoking this method is equivalent to invoking createTempFile(prefix, suffix, null). Returns abstract pathname denoting a newly-created empty file.
28	**public int compareTo(File pathname)** Compares two abstract pathnames lexicographically. Returns zero if the

	argument is equal to this abstract pathname, a value less than zero if this abstract pathname is lexicographically less than the argument, or a value greater than zero if this abstract pathname is lexicographically greater than the argument.
29	**public int compareTo(Object o)** Compares this abstract pathname to another object. Returns zero if the argument is equal to this abstract pathname, a value less than zero if this abstract pathname is lexicographically less than the argument, or a value greater than zero if this abstract pathname is lexicographically greater than the argument.
30	**public boolean equals(Object obj)** Tests this abstract pathname for equality with the given object. Returns true if and only if the argument is not null and is an abstract pathname that denotes the same file or directory as this abstract pathname.

Example

Following is an example to demonstrate File object:

```
package com.tutorialspoint;
import java.io.File;
public class FileDemo {
        public static void main(String[] args) {
              File f = null;
              String[] strs = {"test1.txt", "test2.txt"};
              try{
                  // for each string in string array
                  for(String s:strs )
                  {
                         // create new file
                         f= new File(s);
                         // true if the file is executable
                          boolean bool = f.canExecute();

                          // find the absolute path
                          String a = f.getAbsolutePath();
                           // prints absolute path
                           System.out.print(a);
                            // prints
                            System.out.println(" is executable: "+ bool);
                  }
 }catch(Exception e){
                // if any I/O error occurs
                e.printStackTrace();
        }     }     }
```

Consider there is an executable file test1.txt and another file test2.txt is non executable in the current directory. Let us compile and run the above program, this will produce the following result:

test1.txt is executable: true
test2.txt is executable: false

ii) FileReader Class

- This class inherits from the InputStreamReader class. FileReader is used for reading streams of characters.
- This class has several constructors to create required objects. Following is the list of constructors provided by the FileReader class.

Table: 22. 13

S.No	Constructors and Description
1	**FileReader(File file)** This constructor creates a new FileReader, given the File to read from.
2	**FileReader(FileDescriptor fd)** This constructor creates a new FileReader, given the FileDescriptor to read from.
3	**FileReader(String fileName)** This constructor creates a new FileReader, given the name of the file to read from.

Once you have FileReader object in hand then there is a list of helper methods which can be used to manipulate the files.

Table: 22. 14

S.No	Methods with Description
1	**public int read() throws IOException** Reads a single character. Returns an int, which represents the character read.
2	**public int read(char [] c, int offset, int len)** Reads characters into an array. Returns the number of characters read.

Example

Following is an example to demonstrate class:

```
import java.io.*;
public class FileRead{
    public static void main(String args[])throws IOException{
        File file = new File("Hello1.txt");
        // creates the file
        file.createNewFile();
        // creates a FileWriter Object
        FileWriter writer = new FileWriter(file);
        // Writes the content to the file
        writer.write("This\n is\n an\n example\n");
        writer.flush();
        writer.close();
        //Creates a FileReader Object
        FileReader fr = new FileReader(file);
        char [] a = new char[50];
        fr.read(a); // reads the content to the array
        for(char c : a)
System.out.print(c); //prints the characters one by one
        fr.close();
    }
}
```

This will produce the following result:

```
This
is
an
example
```

iii) FileWriter Class

- This class inherits from the OutputStreamWriter class. The class is used for writing streams of characters.
- This class has several constructors to create required objects. Following is a list.

Table: 22. 15

S.No	Constructors and Description
1	**FileWriter(File file)** This constructor creates a FileWriter object given a File object.
2	**FileWriter(File file, boolean append)** This constructor creates a FileWriter object given a File object with a Boolean indicating whether or not to append the data written.
3	**FileWriter(FileDescriptor fd)** This constructor creates a FileWriter object associated with the given file descriptor.
4	**FileWriter(String fileName)** This constructor creates a FileWriter object, given a file name.
5	**FileWriter(String fileName, boolean append)** This constructor creates a FileWriter object given a file

Once you have FileWriter object in hand, then there is a list of helper methods, which can be used to manipulate the files.

Table: 22. 16

S.No	Methods with Description
1	**public void write(int c) throws IOException** Writes a single character.
2	**public void write(char [] c, int offset, int len)** Writes a portion of an array of characters starting from offset and with a length of len.
3	**public void write(String s, int offset, int len)** Write a portion of a String starting from offset and with a length of len.

Example

Following is an example to demonstrate class:

```
import java.io.*;
public class FileRead{
  public static void main(String args[])throws IOException{
        File file = new File("Hello1.txt");
        // creates the file
        file.createNewFile();
        // creates a FileWriter Object
        FileWriter writer = new FileWriter(file);
        // Writes the content to the file
        writer.write("This\n is\n an\n example\n");
        writer.flush();
        writer.close();
        //Creates a FileReader Object
        FileReader fr = new FileReader(file);
        char [] a = new char[50];
        fr.read(a); // reads the content to the array
        for(char c : a)
                System.out.print(c); //prints the characters one by one
        fr.close();
        }
    }
```

This will produce the following result:

```
This
is
an
example
```

CHAPTER-22

NETWORKING

The term network programming refers to writing programs that execute across multiple devices (computers), in which the devices are all connected to each other using a network.

The java.net package of the J2SE APIs contains a collection of classes and interfaces that provide the low-level communication details, allowing you to write programs that focus on solving the problem at hand.

The java.net package provides support for the two common network protocols:

1. **TCP:** TCP stands for Transmission Control Protocol, which allows for reliable communication between two applications. TCP is typically used over the Internet Protocol, which is referred to as TCP/IP.
2. **UDP:** UDP stands for User Datagram Protocol, a connection-less protocol that allows for packets of data to be transmitted between applications.

This chapter gives a good understanding on the following two subjects:

1. **Socket Programming:** This is the most widely used concept in Networking and it has been explained in very detail.
2. **URL Processing:** This would be covered separately. Click here to learn about URL Processing in Java language.

23.1 URL Processing

- URL stands for Uniform Resource Locator and represents a resource on the World Wide Web, such as a Web page or FTP directory.
- This section shows you how to write Java programs that communicate with a URL. A URL can be broken down into parts, as follows:

protocol://host:port/path?query#ref

Examples of protocols include HTTP, HTTPS, FTP, and File. The path is also referred to as the filename, and the host is also called the authority.

The following is a URL to a web page whose protocol is HTTP:

http://www.amrood.com/index.htm?language=en#j2se

Notice that this URL does not specify a port, in which case the default port for the protocol is used. With HTTP, the default port is 80.

23.2 URL Class Methods

- The java.net.URL class represents a URL and has a complete set of methods to manipulate URL in Java.

The URL class has several constructors for creating URLs, including the following:

S.No	Methods with Description
1	**public URL(String protocol, String host, int port, String file) throws MalformedURLException** Creates a URL by putting together the given parts.
2	**public URL(String protocol, String host, String file) throws MalformedURLException** Identical to the previous constructor, except that the default port for the given protocol is used.
3	**public URL(String url) throws MalformedURLException** Creates a URL from the given String.
4	**public URL(URL context, String url) throws MalformedURLException** Creates a URL by parsing together the URL and String arguments.

Table: 23. 1

The URL class contains many methods for accessing the various parts of the URL being represented. Some of the methods in the URL class include the following:

S.No	Methods with Description
1	**public String getPath()** Returns the path of the URL.
2	**public String getQuery()** Returns the query part of the URL.
3	**public String getAuthority()** Returns the authority of the URL.
4	**public int getPort()** Returns the port of the URL.
5	**public int getDefaultPort()** Returns the default port for the protocol of the URL.
6	**public String getProtocol()** Returns the protocol of the URL.
7	**public String getHost()** Returns the host of the URL
8	**public String getHost()** Returns the host of the URL.
9	**public String getFile()** Returns the filename of the URL.
10	**public String getRef()** Returns the reference part of the URL.
11	**public URLConnection openConnection() throws IOException** Opens a connection to the URL, allowing a client to communicate with the resource.

Table: 23. 2

Example

The following URLDemo program demonstrates the various parts of a URL. A URL is entered on the command line, and the URLDemo program outputs each part of the given URL.

```
// File Name : URLDemo.java
import java.net.*;
import java.io.*;
public class URLDemo
{
        public static void main(String [] args)
        {
            try
            {
 URL url = new URL("http://www.amrood.com/index.htm?language=en#j2se");

 System.out.println("URL is " + url.toString());
 System.out.println("protocol is " + url.getProtocol());
 System.out.println("authority is " + url.getAuthority());
 System.out.println("file name is " + url.getFile());
 System.out.println("host is " + url.getHost());
 System.out.println("path is " + url.getPath());
 System.out.println("port is " + url.getPort());
 System.out.println("default port is " + url.getDefaultPort());
 System.out.println("query is " + url.getQuery());
 System.out.println("ref is " + url.getRef());
 }catch(IOException e)
 {
          e.printStackTrace();
      }
   }
}
```

A sample run of this program will produce the following result:

```
URL is http://www.amrood.com/index.htm?language=en#j2se
protocol is http
authority is www.amrood.com
file name is /index.htm?language=en
host is www.amrood.com
path is /index.htm
port is -1
default port is 80
query is language=en
ref is j2se
```

23.3 URLConnections Class Methods

- The openConnection() method returns a java.net.URLConnection, an abstract class whose subclasses represent the various types of URL connections.

For example:

- If you connect to a URL whose protocol is HTTP, the openConnection() method returns an HttpURLConnection object.
- If you connect to a URL that represents a JAR file, the openConnection() method returns a JarURLConnection object, etc.

The URLConnection class has many methods for setting or determining information about the connection, including the following:

Table: 23. 3

S.No	Methods with Description
1	**Object getContent()** Retrieves the contents of this URL connection.
2	**Object getContent(Class[] classes)** Retrieves the contents of this URL connection.
3	**String getContentEncoding()** Returns the value of the content-encoding header field.
4	**int getContentLength()** Returns the value of the content-length header field.
5	**String getContentType()** Returns the value of the content-type header field.
6	**int getLastModified()** Returns the value of the last-modified header field.

7	**long getExpiration()** Returns the value of the expired header field
8	**long getIfModifiedSince()** Returns the value of this object's ifModifiedSince field.
9	**public void setDoInput(boolean input)** Passes in true to denote that the connection will be used for input. The default value is true because clients typically read from a URLConnection.
10	**public void setDoOutput(boolean output)** Passes in true to denote that the connection will be used for output. The default value is false because many types of URLs do not support being written to.
11	**public InputStream getInputStream() throws IOException** Returns the input stream of the URL connection for reading from the resource.
12	**public OutputStream getOutputStream() throws IOException** Returns the output stream of the URL connection for writing to the resource.
13	**public URL getURL()** Returns the URL that this URLConnection object is connected to.

Example

The following URLConnectionDemo program connects to a URL entered from the command line.

If the URL represents an HTTP resource, the connection is cast to HttpURLConnection, and the data in the resource is read one line at a time.

```
// File Name : URLConnDemo.java
import java.net.*;
import java.io.*;
public class URLConnDemo
{
    public static void main(String [] args)
{
    try
    {
        URL url = new URL("http://www.amrood.com");
        URLConnection urlConnection = url.openConnection();
        HttpURLConnection connection = null;
        if(urlConnection instanceof HttpURLConnection)
        {
            connection = (HttpURLConnection) urlConnection;
```

```
        }
        else
        {
            System.out.println("Please enter an HTTP URL.");
            return;
        }
        BufferedReader in = new BufferedReader(
        new InputStreamReader(connection.getInputStream()));
          String urlString = "";
          String current;
          while((current = in.readLine()) != null)
          {
             urlString += current;
          }
          System.out.println(urlString);
    }catch(IOException e)
    {
           e.printStackTrace();
    }
    }
    }
```

A sample run of this program will produce the following result:

```
$ java URLConnDemo

.....a complete HTML content of home page of amrood.com.....
```

23.4 Socket Programming

- Sockets provide the communication mechanism between two computers using TCP. A client program creates a socket on its end of the communication and attempts to connect that socket to a server.
- When the connection is made, the server creates a socket object on its end of the communication. The client and the server can now communicate by writing to and reading from the socket.
- The java.net.Socket class represents a socket, and the java.net.ServerSocket class provides a mechanism for the server program to listen for clients and establish connections with them.

The following steps occur when establishing a TCP connection between two computers using sockets:

- The server instantiates a ServerSocket object, denoting which port number communication is to occur on.
- The server invokes the accept() method of the ServerSocket class. This method waits until a client connects to the server on the given port.
- After the server is waiting, a client instantiates a Socket object, specifying the server name and the port number to connect to.
- The constructor of the Socket class attempts to connect the client to the specified server and the port number. If communication is established, the client now has a Socket object capable of communicating with the server.
- On the server side, the accept() method returns a reference to a new socket on the server that is connected to the client's socket.

After the connections are established, communication can occur using I/O streams. Each socket has both an OutputStream and an InputStream. The client's OutputStream is connected to the server's InputStream, and the client's InputStream is connected to the server's OutputStream.

TCP is a two-way communication protocol, hence data can be sent across both streams at the same time. Following are the useful classes providing complete set of methods to implement sockets.

23.5 ServerSocket Class Methods

The **java.net.ServerSocket** class is used by server applications to obtain a port and listen for client requests.

The ServerSocket class has four constructors:

Table: 23. 4

S.No	Methods with Description
1	**public ServerSocket(int port) throws IOException** Attempts to create a server socket bound to the specified port. An exception occurs if the port is already bound by another application.
2	**public ServerSocket(int port, int backlog) throws IOException** Similar to the previous constructor, the backlog parameter specifies how many incoming clients to store in a wait queue.
3	**public ServerSocket(int port, int backlog, InetAddress address) throws IOException** Similar to the previous constructor, the InetAddress parameter specifies the local IP address to bind to. The InetAddress is used for servers that may have multiple IP addresses, allowing the server to specify which of its IP addresses to accept client requests on.
4	**public ServerSocket() throws IOException** Creates an unbound server socket. When using this constructor, use the bind() method when you are ready to bind the server socket.

If the ServerSocket constructor does not throw an exception, it means that your application has successfully bound to the specified port and is ready for client requests.

Following are some of the common methods of the ServerSocket class:

Table: 23. 5

S.No	Methods with Description
1	**public int getLocalPort()** Returns the port that the server socket is listening on. This method is useful if you passed in 0 as the port number in a constructor and let the server find a port for you.
2	**public Socket accept() throws IOException** Waits for an incoming client. This method blocks until either a client connects to the server on the specified port or the socket times out, assuming that the time-out value has been set using the setSoTimeout() method. Otherwise, this method blocks indefinitely.
3	**public void setSoTimeout(int timeout)** Sets the time-out value for how long the server socket waits for a client during the accept().
4	**public void bind(SocketAddress host, int backlog)** Binds the socket to the specified server and port in the SocketAddress object. Use this method if you have instantiated the ServerSocket using the noargument constructor.

When the ServerSocket invokes accept(), the method does not return until a client connects. After a client does connect, the ServerSocket creates a new Socket on an unspecified port and returns a reference to this new Socket. A TCP connection now exists between the client and the server, and communication can begin.

23.6 Socket Class Methods

- The **java.net.Socket** class represents the socket that both the client and the server use to communicate with each other.
- The client obtains a Socket object by instantiating one, whereas the server obtains a Socket object from the return value of the accept() method.

The Socket class has five constructors that a client uses to connect to a server:

Table: 23. 6

S.No	Methods with Description
1	**public Socket(String host, int port) throws UnknownHostException, IOException.** This method attempts to connect to the specified server at the specified port. If this constructor does not throw an exception, the connection is successful and the client is connected to the server.

2	**public Socket(InetAddress host, int port) throws IOException** This method is identical to the previous constructor, except that the host is denoted by an InetAddress object.
3	**public Socket(String host, int port, InetAddress localAddress, int localPort) throws IOException.** Connects to the specified host and port, creating a socket on the local host at the specified address and port.
4	**public Socket(InetAddress host, int port, InetAddress localAddress, int localPort) throws IOException.** This method is identical to the previous constructor, except that the host is denoted by an InetAddress object instead of a String.
5	**public Socket()** Creates an unconnected socket. Use the connect() method to connect this socket to a server.

When the Socket constructor returns, it does not simply instantiate a Socket object but it actually attempts to connect to the specified server and port. Some methods of interest in the Socket class are listed here. Notice that both the client and the server have a Socket object, so these methods can be invoked by both the client and the server.

Table: 23. 7

S.No	Methods with Description
1	**public void connect(SocketAddress host, int timeout) throws IOException** This method connects the socket to the specified host. This method is needed only when you instantiate the Socket using the no-argument constructor.
2	**public InetAddress getInetAddress()** This method returns the address of the other computer that this socket is connected to.
3	**public int getPort()** Returns the port the socket is bound to on the remote machine.
4	**public int getLocalPort()** Returns the port the socket is bound to on the local machine.
5	**public SocketAddress getRemoteSocketAddress()** Returns the address of the remote socket.
6	**public InputStream getInputStream() throws IOException** Returns the input stream of the socket. The input stream is connected to the output stream of the remote socket.
7	**public OutputStream getOutputStream() throws IOException** Returns the output stream of the socket. The output stream is connected to the input stream of the remote socket.
8	**public void close() throws IOException** Closes the socket, which makes this Socket object no longer capable of connecting again to any server.

23.7 INETAddress Class Methods

This class represents an Internet Protocol (IP) address. Here are following usefull methods which you would need while doing socket programming:

Table: 23. 8

S.No	Methods with Description
1	**static InetAddress getByAddress(byte[] addr)** Returns an InetAddress object given the raw IP address.
2	**static InetAddress getByAddress(String host, byte[] addr)** Creates an InetAddress based on the provided host name and IP address.
3	**static InetAddress getByName(String host)** Determines the IP address of a host, given the host's name.
4	**String getHostAddress()** Returns the IP address string in textual presentation.
5	**String getHostName()** Gets the host name for this IP address.
6	**static InetAddress InetAddress getLocalHost()** Returns the local host.
7	**String toString()** Converts this IP address to a String.

Socket Client Example:

The following GreetingClient is a client program that connects to a server by using a socket and sends a greeting, and then waits for a response.

```
// File Name GreetingClient.java
import java.net.*;
import java.io.*;
public class GreetingClient
{
 public static void main(String [] args)
 {
 String serverName = args[0];
 int port = Integer.parseInt(args[1]);
 try
 {
 System.out.println("Connecting to " + serverName +
" on port " + port);
 Socket client = new Socket(serverName, port);
 System.out.println("Just connected to "
+ client.getRemoteSocketAddress());
 OutputStream outToServer = client.getOutputStream();
 DataOutputStream out = new DataOutputStream(outToServer);
 out.writeUTF("Hello from "
 + client.getLocalSocketAddress());
 InputStream inFromServer = client.getInputStream();
 DataInputStream in =
 new DataInputStream(inFromServer);
 System.out.println("Server says " + in.readUTF());
 client.close();
 }catch(IOException e)
 {
 e.printStackTrace();
 }
 }
}
```

Socket Server Example:

The following Greeting Server program is an example of a server application that uses the Socket class to listen for clients on a port number specified by a command-line argument:

```
// File Name GreetingServer.java
import java.net.*;
import java.io.*;
public class GreetingServer extends Thread
{
 private ServerSocket serverSocket;

 public GreetingServer(int port) throws IOException
 {
 serverSocket = new ServerSocket(port);
 serverSocket.setSoTimeout(10000);
 }
 public void run()
 {
 while(true)
 {
 try
 {
 System.out.println("Waiting for client on port " +
 serverSocket.getLocalPort() + "...");
 Socket server = serverSocket.accept();
 System.out.println("Just connected to "
 + server.getRemoteSocketAddress());
DataInputStream in = new DataInputStream(server.getInputStream());
 System.out.println(in.readUTF());
DataOutputStream out =
 new DataOutputStream(server.getOutputStream());
 out.writeUTF("Thank you for connecting to "
 + server.getLocalSocketAddress() + "\nGoodbye!");
 server.close();
 }catch(SocketTimeoutException s)
```

Conti........

```
{
System.out.println("Socket timed out!");
break;
}catch(IOException e)
{
e.printStackTrace();
break;
}
}
}
public static void main(String [] args)
{
int port = Integer.parseInt(args[0]);
try
{
Thread t = new GreetingServer(port);
t.start();
}catch(IOException e)
{
e.printStackTrace();
}
}
}
```

Compile the client and the server and then start the server as follows:

```
$ java GreetingServer 6066
Waiting for client on port 6066...
```

Check the client program as follows:

```
$ java GreetingClient localhost 6066
Connecting to localhost on port 6066
Just connected to localhost/127.0.0.1:6066
Server says Thank you for connecting to /127.0.0.1:6066
Goodbye!
```

23.8 Event Handling

What is an Event?

- Change in the state of an object is known as event i.e. event describes the change in state of source.
- Events are generated as result of user interaction with the graphical user interface components.
- For example, clicking on a button, moving the mouse, entering a character through keyboard, selecting an item from list, scrolling the page are the activities that causes an event to happen.

Types of Event

The events can be broadly classified into two categories:

Foreground Events:

- Those events which require the direct interaction of users.
- They are generated as consequences of a person interacting with the graphical components in Graphical User Interface.
- For example, clicking on a button, moving the mouse, entering a character through keyboard, selecting an item from list, scrolling the page etc.

Background Events:

- Those events that require the interaction of end user are known as background events.
- Operating system interrupts, hardware or software failure, timer expires, an operation completion are the example of background events.

What is Event Handling?

- Event Handling is the mechanism that controls the event and decides what should happen if an event occurs.
- This mechanism has the code which is known as event handler that is executed when an event occurs. Java Uses the Delegation Event Model to handle the events.
- This model defines the standard mechanism to generate and handle the events. Let's have a brief introduction to this model.

The Delegation Event Model has the following key participants namely:

Source:

- The source is an object on which event occurs. Source is responsible for providing information of the occurred event to its handler.
- Java provide as with classes for source object.

Listener:

- It is also known as event handler.
- Listener is responsible for generating response to an event. From java implementation point of view the listener is also an object.
- Listener waits until it receives an event. Once the event is received, the listener processes the event a then returns.

The benefit of this approach is that the user interface logic is completely separated from the logic that generates the event. The user interface element is able to delegate the processing of an event to the separate piece of code. In this model, Listener needs to be registered with the source object so that the listener can receive the event notification. This is an efficient way of handling the event because the event notifications are sent only to those listeners that want to receive them.

23.9 Steps Involved In Event Handling

- The User clicks the button and the event is generated.
- Now the object of concerned event class is created automatically and information about the source and the event get populated with in same object.
- Event object is forwarded to the method of registered listener class.
- The method is now get executed and returns.

Points to remember about listener:

- In order to design a listener class we have to develop some listener interfaces. These Listener interfaces forecast some public abstract call-back methods which must be implemented by the listener class.
- If you do not implement the any if the predefined interfaces then your class can not act as a listener class for a source object.

23.10 Important Event Classes and Interface

Table: 23. 9

Event Classes	Description	Listener Interface
ActionEvent	generated when button is pressed, menu-item is selected, list-item is double clicked	ActionListener
MouseEvent	generated when mouse is dragged, moved, clicked, pressed or released and also when it enters or exit a component	MouseListener
KeyEvent	generated when input is received from keyboard	KeyListener
ItemEvent	generated when check-box or list item is clicked	ItemListener
TextEvent	generated when value of textarea or textfield is changed	TextListener
MouseWheelEvent	generated when mouse wheel is moved	MouseWheelListener
WindowEvent	generated when window is activated, deactivated, deiconified, iconified, opened or closed	WindowListener
ComponentEvent	generated when component is hidden, moved, resized or set visible	ComponentEventListener

ContainerEvent	generated when component is added or removed from container	ContainerListener
AdjustmentEvent	generated when scroll bar is manipulated	AdjustmentListener
FocusEvent	generated when component gains or loses keyboard focus	FocusListener

Example of Event Handling:

```
import java.awt.*;
import java.awt.event.*;
import java.applet.*;
import java.applet.*;
import java.awt.event.*;
import java.awt.*;
public class Test extends Applet implements KeyListener      {
String msg="";
public void init()
 {
  addKeyListener(this);
 }
 public void keyPressed(KeyEvent k)
 {
  showStatus("KeyPressed");
 }
```

Conti......

```
public void keyReleased(KeyEvent k)
 {
  showStatus("KeyRealesed");
 }
 public void keyTyped(KeyEvent k)
 {
  msg = msg+k.getKeyChar();
  repaint();
 }
 public void paint(Graphics g)
 {
  g.drawString(msg, 20, 40);
 }
}
```

CHAPTER-23

AWT

24.1Overview of the AWT

AWT stands for Abstract Window ToolKit. It is a portable GUI library between Solaris and Windows 95/NT and Mac System 7.X(soon) for stand-alone applications and/or applets. Since it can be used in applets it can be used on IRIX, SunOS, HP/UX, and Linux which Netscape 2.0 supports.

The Abstract Window Toolkit provides many classes for programmers to use. It is your connection between your application and the native GUI. The AWT hides you from the underlying details of the GUI your application will be running on and thus is at very high level of abstraction. It takes the lowest common denominator approach to retain portability. No floating toolbars or Balloon helps here...

It is a Java package and can be used in any Java program by importing **java.awt.*** via the import keyword. The documentation for the package is available at the Java homepage. The package will be covered briefly as this document is not considered advanced material because it does not discuss Peers, Image Consumers/Producers, Toolkits and other advanced AWT ilk. It is recommended you look at the source code to see how the AWT really works.

24.2 Structure of the AWT

The structure of the AWT is rather simple: Components are added to and then laid out by layout managers in Containers. We have a variety of event handling, menu, fonts, graphics classes in addition to those two, but they'll be only briefly discussed here.

Nothing prevents us from using threads, audio, I/O, networking classes alongside the AWT. The AWT by itself is rather bare and empty. Imagine a text editor that couldn't save! As a side note, the AWT can be used in a stand-alone Java application or in an applet that is embedded within a HTML document.

24.3 Working with AWT Windows

The AWT contains numerous classes and methods that allow us to create and manage windows. Although the main purpose of the AWT is to support applet windows, it can also be used to create stand -alone windows that run in a GUI environment, such as Windows.

i) AWT Classes

The AWT classes are contained in the **java.awt** package. It is one of Java's largest packages. Fortunately, because it is logically organized in a top down, hierarchical fashion, it is easier to understand and use than you might at first believe.

ii) Window Fundamentals

The AWT defines windows according to a class hierarchy that adds functionality and specificity with each level. The two most common windows are those derived from Panel, which is used by applets, and those derived from Frame, which creates a standard window. Much of the functionality of these windows is derived from their parent classes. Thus, a description of the class hierarchies relating to these two classes is fundamental to their understanding. Figure below shows the class hierarchy for Panel and Frame.

24.4 Applet Class Hierarchy

i) Component

At the top of the AWT hierarchy is the Component class. Component is an abstract class that encapsulates all of the attributes of a visual component. All user interface elements that are displayed on the screen and that interact with the user are subclasses of Component. It defines over a hundred public methods that are responsible for managing events, such as mouse and keyboard input, positioning and sizing the window, and repainting. A Component object is responsible for remembering the current foreground and background colors and the currently selected text font.

ii) Container

The Container class is a subclass of Component. It has additional methods that allow other Component objects to be nested within it. Other Container objects can be stored inside of a Container (since they are themselves instances of Component). This makes for a multileveled containment system. A container is responsible for laying out (that is, positioning) any components that it contains. It does this through the use of various layout managers.

iii) Panel

The Panel class is a concrete subclass of Container. It doesn't add any new methods; it simply implements Container. A Panel may be thought of as a recursively nestable, concrete screen component. Panel is the super-class for Applet. When screen output is directed to an applet, it is drawn on the surface of a Panel object. In essence, a Panel is a window that does not contain a title bar, menu bar, or border. This is why we don't see these items when an applet is run inside a browser. When we run an applet using an applet viewer, the applet viewer provides the title and border. Other components can be added to a Panel object by its add( ) method (inherited from Container). Once these components have been added, we can position and resize them manually using the setLocation( ), setSize( ), or setBounds() methods defined by Component.

iv)Window

The Window class creates a top-level window. A top -level window is not contained within any other object; it sits directly on the desktop. Generally, we won't create Window objects directly. Instead, we will use a subclass of Window called Frame.

v) Frame

Frame encapsulates what is commonly thought of as a "window." It is a subclass of Window and has a title bar, menu bar, borders, and resizing corners. If we create a Frame object from within an applet, it will contain a warning message, such as "Java Applet Window," to the user that an applet window has been created. This message warns users that the window they see was started by an applet and not by software running on their computer. When a Frame window is created by a program rather than an applet, a normal window is created.

vi) Canvas

Canvas encapsulates a blank window upon which we can draw.

vii) Dimension

This class encapsulates the 'width' and 'height' of a component (in integer precision) in a single object. The class is associated with certain properties of components. Several methods defined by the Component class and the Layout Manager Interface return a Dimension object.

Normally the values of width and height are non -negative integers. The constructors that allow us to create a dimension do not prevent us from setting a negative value for these properties. If the value of width or height is negative, the behaviour of some methods defined by other objects is undefined.

Fields of Dimension:

int height	**The height dimension; negative values can be used.**
int width	**The width dimension; negative values can be used.**

24.5 Constructors

Dimension()

It creates an instance of Dimension with a width of zero and a height of zero.

Dimension(Dimension d)

It creates an instance of Dimension whose width and height are the same as for the specified dimension.

Dimension(int width, int height)

It constructs a Dimension and initializes it to the specified width and specified height.

Methods:

boolean equals(Object obj)

It checks whether two dimension objects have equal values.

double getHeight()

It returns the height of this dimension in double precision.

Dimension getSize()

It gets the size of this Dimension object.

```
double getWidth()
```

It returns the width of this dimension in double precision.

```
void setSize(Dimension d)
```

It sets the size of this Dimension object to the specified size.

```
void setSize(double width, double height)
```

It sets the size of this Dimension object to the specified width and height in double precision.

```
void setSize(int width, int height)
```

It sets the size of this Dimension object to the specified width and height.

24.6 Working with Frame Windows

After the applet, the type of window we will most often create is derived from Frame. We will use it to create child windows within applets, and top-level or child windows for applications. It creates a standard -style window. Following are two of Frame's constructors:

Frame( ) Frame(String title)

The first form creates a standard window that does not contain a title. The second form creates a window with the title specified by title. Note that we cannot specify the dimensions of the window. Instead, we must set the size of the window after it has been created.

Setting the Windows Dimensions

The setSize( ) method is used to set the dimensions of the window. Its signature is shown below:

```
void setSize(int newWidth, int newHeight)

void setSize(Dimension newSize)
```

The new size of the window is specified by 'newWidth' and 'newHeight', or by the 'width' and 'height' fields of the Dimension object passed in 'newSize'. The dimensions are specified in terms of pixels. The getSize( ) method is used to obtain the current size of a window. Its signature is:

```
Dimension getSize( )
```

This method returns the current size of the window contained within the 'width' and 'height' fields of a Dimension object.

24.7 AWT Controls

Every user interface considers the following three main aspects:

- **UI elements:** These are the core visual elements the user eventually sees and interacts with. GWT provides a huge list of widely used and common elements varying from basic to complex.
- **Layouts:** They define how UI elements should be organized on the screen and provide a final look and feel to the GUI (Graphical User Interface).
- **Behavior:** These are events which occur when the user interacts with UI elements.

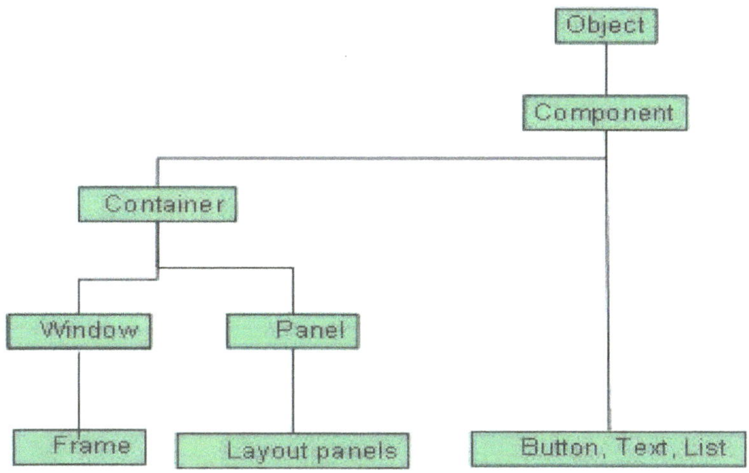

Fig: 24.1

24.8 AWT UI Elements:

Following is the list of commonly used controls while designed GUI using AWT.

Table: 24.1

S.No	Control & Description
1	**Label-**A Label object is a component for placing text in a container.
2	**Button-**This class creates a labeled button.
3	**Check box-**A check box is a graphical component that can be in either an on (true) or off (false) state.
4	**CheckboxGroup-**The CheckboxGroup class is used to group the set of checkbox.
5	**List-**The List component presents the user with a scrolling list of text items.
6	**TextField-**A TextField object is a text component that allows for the editing of a single line of text.

7	**TextArea-**A TextArea object is a text component that allows for the editing of a multiple lines of text.
8	**Choice** -A Choice control is used to show popup menu of choices. Selected choice is shown on the top of the menu.
9	**Canvas** -A Canvas control represents a rectangular area where application can draw something or can receive inputs created by user.
10	**Image** -An Image control is superclass for all image classes representing graphical images.
11	**Scrollbar** -A Scrollbar control represents a scroll bar component in order to enable user to select from range of values.
12	**Dialog** -A Dialog control represents a top-level window with a title and a border used to take some form of input from the user.
13	**FileDialog** -A FileDialog control represents a dialog window from which the user can select a file.

Adding and Removing Controls

In order to include a control in a window, we must add it to the window. So, we must first create an instance of the desired control and then add it to a window by calling add( ), which is defined by Container. The add( ) method has several forms.

```
Component add(Component compObj)
```

Here, compObj is an instance of the control that we want to add. A reference to compObj is returned. Once a control has been added, it will automatically be visible whenever its parent window is displayed. Sometimes we will want to remove a control from a window when the control is no longer needed. For doing this, call remove( ). This method is also defined by Container. It has this general form:

```
Void remove(Component obj)
```

Here, obj is a reference to the control that we want to remove. We can remove all controls by calling removeAll( ).

Example:

```
// Demonstrate Labels
import java.awt.*;
import java.applet.*;
public class LabelDemo extends Applet
{
public void init()
{
Label one = new Label("One");
Label two = new Label("Two");
Label three = new Label("Three");
// add labels to applet window
add(one);
add(two);
add(three);
}
}
```

24.9 AWT Layout

The layout manager automatically positions all the components within the container. If we do not use layout manager then also the components are positioned by the default layout manager. It is possible to layout the controls by hand but it becomes very difficult because of the following two reasons.

- It is very tedious to handle a large number of controls within the container.
- Oftenly the width and height information of a component is not given when we need to arrange them.

Java provides us with various layout managers to position the controls. The properties like size, shape and arrangement varies from one layout manager to other layout manager. When the size of the applet or the application window changes the size, shape and arrangement of the components also changes in response i.e. the layout managers adapt to the dimensions of appletviewer or the application window.

The layout manager is associated with every Container object. Each layout manager is an object of the class that implements the LayoutManager interface.

Following are the interfaces defining functionalities of Layout Managers.

Table: 24.2

S.No	Interface & Description
1	**LayoutManager-**The LayoutManager interface declares those methods which need to be implemented by the class whose object will act as a layout manager.
2	**LayoutManager2-**The LayoutManager2 is the sub-interface of the LayoutManager.This interface is for those classes that know how to layout containers based on layout constraint object.

AWT Layout Manager Classes

S.No	LayoutManager & Description
1	**Borderlayout-** The borderlayout arranges the components to fit in the five regions: east, west, north, south and center.
2	**CardLayout-** The CardLayout object treats each component in the container as a card. Only one card is visible at a time.
3	**FlowLayout-** The FlowLayout is the default layout.It layouts the components in a directional flow.
4	**GridLayout-** The GridLayout manages the components in form of a rectangular grid.
5	**GridBagLayout-** This is the most flexible layout manager class. The object of GridBagLayout aligns the component vertically, horizontally or along their baseline without requiring the components of same size.

Table: 24. 3

Example:

```
import java.awt.*;
import javax.swing.*;

public class Border {
JFrame f;
Border(){
   f=new JFrame();     JButton b1=new JButton("NORTH");;
   JButton b2=new JButton("SOUTH");;
   JButton b3=new JButton("EAST");;
   JButton b4=new JButton("WEST");;
   JButton b5=new JButton("CENTER");;

   f.add(b1,BorderLayout.NORTH);
   f.add(b2,BorderLayout.SOUTH);
   f.add(b3,BorderLayout.EAST);
   f.add(b4,BorderLayout.WEST);
       f.add(b5,BorderLayout.CENTER);

   f.setSize(300,300);
   f.setVisible(true);
}
public static void main(String[] args) {
   new Border();
}
}
```

24.10 AWT MENU

This menu bar consists of various menu choices available to the end user. Further each choice contains list of options which is called drop down menus. Menu and MenuItem controls are subclass of MenuComponent class.

Menu Hiearchy

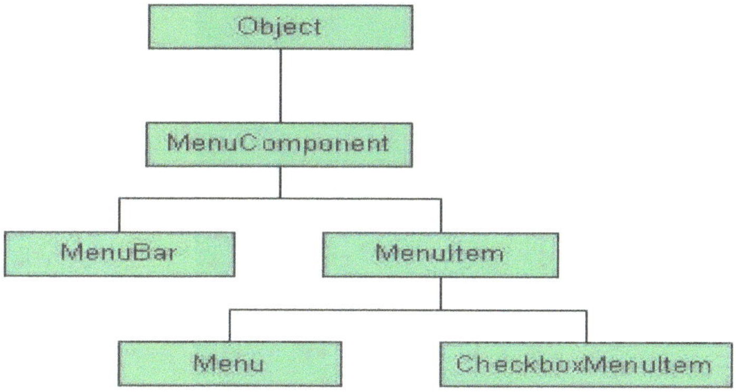

Fig: 24.2

Menu Controls

Table: 24. 4

S.No	Control & Description
1	**Menu Component-** It is the top level class for all menu related controls.
2	**MenuBar-** The MenuBar object is associated with the top-level window.
3	**MenuItem-** The items in the menu must belong to the MenuItem or any of its subclass.
4	**Menu-** The Menu object is a pull-down menu component which is displayed from the menu bar.
5	**CheckboxMenuItem-** CheckboxMenuItem is subclass of MenuItem.
6	**PopupMenu-** PopupMenu can be dynamically popped up at a specified position within a component.

Example:

```
import java.awt.*;
import java.io.*;
import java.awt.event.*;
public class MenuDemo extends Frame implements ActionListener
{
  TextArea ta;
  public MenuDemo()
  {                  // create menu bar
    MenuBar mBar = new MenuBar();
    setMenuBar(mBar);   // add menu bar to frame
                        // create menus
    Menu files = new Menu("Files");
    Menu date = new Menu("Date");
    Menu exit = new Menu("Exit");
    ta = new TextArea(10, 40);
    ta.setBackground(Color.cyan);
                        // add menus to menu bar
    mBar.add(files);
    mBar.add(date);
    mBar.add(exit);
                        // create menu items to menus
    MenuItem mi1 = new MenuItem("GLDemo.java");
    files.add(mi1);              // using anonymous menu items
    files.add(new MenuItem("LabelTest.java"));
    files.addSeparator();
    files.add(new MenuItem("UsingFonts.class"));
    files.add(new MenuItem("FLDemo.class"));

    date.add(new MenuItem("Today"));
    exit.add(new MenuItem("Close"));
                        // linking listener
    files.addActionListener(this);
    date.addActionListener(this);
    exit.addActionListener(this);
  add(ta, "Center");
setTitle("Menu Practice");
    setSize(400, 400);
    setVisible(true);
  }
```

Conti.....

```
 public void actionPerformed(ActionEvent e)
 {
  String str = e.getActionCommand();

  if(str.equals("Close"))
  {
   System.exit(0);
else if(str.equals("Today"))
  {
   ta.setText("Today: " + new java.util.Date());
  }
  else
  {
   try
   {
    FileReader fr = new FileReader(str);
    ta.setText("Folloiwing are file contents:\n\n");
    int temp;
    while( (temp = fr.read()) != -1)
    {
     char ch = (char) temp;
     String s1 = String.valueOf(ch);
     ta.append(s1);
    }
    fr.close();
   }
   catch(IOException e1)
   {
    ta.setText("Exception: " + e1);
   }
  }
 }
 public static void main(String args[])
 {
  new MenuDemo();
 }
}
```

24.11 AWT SWING

Java Swing tutorial is a part of **Java Foundation Classes (JFC)** that is used to create window-based applications. It is built on the top of **AWT (Abstract Windowing Toolkit)** API and entirely written in java.

Unlike AWT, Java Swing provides platform-independent and lightweight components.

The javax.swing package provides classes for java swing API such as **JButton, JTextField, JTextArea, JRadioButton, JCheckbox, JMenu, JColorChooser** etc.

Difference between AWT and Swing

There are many differences between java AWT and swing that are given below.

S.No	JAVA AWT	JAVA SWING
1	AWT components are **platform-dependent.**	Java swing components are **platform-independent.**
2	AWT components are **heavyweight.**	Swing components are **lightweight.**
3	AWT **doesn't support pluggable look and feel.**	Swing **supports pluggable look and feel.**
4	AWT provides **less components** than Swing.	Swing provides **more powerful componentssuch** as tables, lists, scrollpanes, colorchooser, tabbedpane etc.
5	AWT **doesn't follows MVC**(Model View Controller) where model represents data, view represents presentation and controller acts as an interface between model and view.	Swing **follows MVC.**

Table: 24.5

Hierarchy of Java Swing classes

The hierarchy of java swing API is given below.

Fig: 24.3

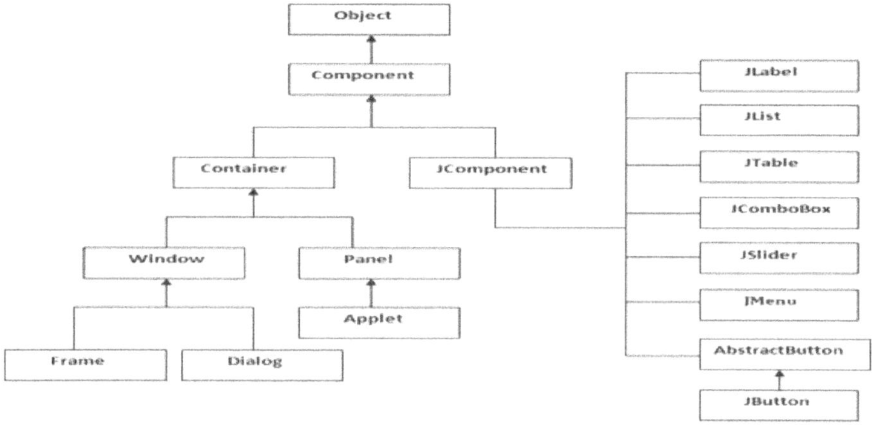

Commonly used Methods of Component class

Table: 24. 6

Method	Description
public void add(Component c)	add a component on another component.
public void setSize(int width,int height)	sets size of the component.
public void setLayout(LayoutManager m)	sets the layout manager for the component.
public void setVisible(boolean b)	sets the visibility of the component. It is by default false.

Java Swing Examples

There are two ways to create a frame:

- By creating the object of Frame class (association)
- By extending Frame class (inheritance)
- We can write the code of swing inside the main(), constructor or any other method.

Simple Java Swing Example

Let's see a simple swing example where we are creating one button and adding it on the JFrame object inside the main() method.

File: FirstSwingExample.java

```
import javax.swing.*;
public class FirstSwingExample {
public static void main(String[] args) {
JFrame f=new JFrame();//creating instance of JFrame
JButton b=new JButton("click");//creating instance of JButton
b.setBounds(130,100,100, 40);//x axis, y axis, width, height
f.add(b);//adding button in JFrame
f.setSize(400,500);//400 width and 500 height
f.setLayout(null);//using no layout managers
f.setVisible(true);//making the frame visible
}
}
```

Example of Swing by Association inside constructor

We can also write all the codes of creating JFrame, JButton and method call inside the java constructor.

File: Simple.java

```
import javax.swing.*;
public class Simple {
JFrame f;
Simple(){
f=new JFrame();//creating instance of JFrame

JButton b=new JButton("click");//creating instance of JButton
b.setBounds(130,100,100, 40);
 f.add(b);//adding button in JFrame
f.setSize(400,500);//400 width and 500 height
f.setLayout(null);//using no layout managers
f.setVisible(true);//making the frame visible
}
  public static void main(String[] args) {
new Simple();
}
}
```

The setBounds(int xaxis, int yaxis, int width, int height)is used in the above example that sets the position of the button.

Simple Example of Swing by Inheritance

We can also inherit the JFrame class, so there is no need to create the instance of JFrame class explicitly.

File: Simple2.java

```
import javax.swing.*;
public class Simple2 extends JFrame{//inheriting JFrame
JFrame f;
Simple2(){
JButton b=new JButton("click");//create button
b.setBounds(130,100,100, 40);
add(b);//adding button on frame
setSize(400,500);
setLayout(null);
setVisible(true);
}
public static void main(String[] args) {
new Simple2();
}
}
```

CHAPTER- 24

MULTIMEDIA

25.1 Introduction to Java Multimedia Image Animation

- Multimedia is becoming inevitable for a common man today or to a programmer. The term multimedia includes the integration of many aspects like audio, video, graphics, images and animation.
- Multimedia recorded once can be played any number of times at the user's choice. For this, the computer system should come with suitable hardware like video graphics adapter (VGA), high speed microprocessor and RAM etc.
- Multimedia can be linear or non-linear. With linear multimedia, user cannot navigate and with non-linear, user can navigate wherever he would like as in games.
- Images (derived from imago, a Latin word) can be captured and fed to a computer with versatile instruments like cell phones, camera and telescope etc.

25.2 Image and Animation

Java Image Animation Basics

- One of the features of Java is to support image drawing and animation.
- The image size depends on the actual size (how much it is big), transparency and its color depth.
- A few popular image formats are gif (Graphics Interchange Format), jpeg (Joint Photographic Experts Group), bmp (Bit-mapped graphics format), TIFF (Tagged Image File Format), PNG (Portable Network Graphics) and WBMP (Wireless BitMap) etc.
- Java, from JDK 1.5, supports all these formats. Every format includes its own style of arranging the pixels. An image size is measured in terms of bytes.
- A format comes with its own style of placing the pixels, in rows and columns, in a systematic way. Images take a lot of space that too images taken with high-resolution cameras.

Both Java.awt and Java.swing Packages Display Images

```
import java.applet.Applet;
import java.awt.*;
import javax.swing.*;
public class LoadImageAndScale extends JApplet {
private Image logo1;
private ImageIcon logo2;
// load the image when the applet is loaded
public void init()
{
logo1 = getImage( getDocumentBase(), "logo2.gif" );
logo2 = new ImageIcon( "logo2.gif" );
}
{
 // draw the original image
g.drawImage( logo1, 0, 0, this );
// draw the image scaled to fit the width of the applet
// and the height of the applet minus 120 pixels
g.drawImage( logo1, 0, 120,
getWidth(), getHeight() - 120, this );
// draw the icon using its paintIcon method
logo2.paintIcon( this, g, 180, 0 );
}
}
## HTML Code
<html>
<applet code="LoadImageAndScale.class" width=340 height=340>
</applet>
</html>
```

25.3 Animating a Series of Images

- When you play some video games, you are not aware that the program is just using a number of images and animating them to give the appearance of a moving picture, movies and cartoons all use this method, just a numbers of pictures linked together and viewed at high speed.
- The problems that you face when trying to create moving images is that the end result does not play smoothly, they appear choppy or the characters do not appear to move properly.

The below example is a very basic animation program, there are complete gaming/graphics books on how to program animation in Java, explaining the problems of animating images and offering solutions.

```
import java.awt.*;
import java.awt.event.*;
import javax.swing.*;
public class LogoAnimator extends JPanel implements ActionListener {
protected ImageIcon images[];
protected int totalImages = 30,
currentImage = 0,
animationDelay = 50; // 50 millisecond delay
protected Timer animationTimer;
public LogoAnimator()
{
setSize( getPreferredSize() );
images = new ImageIcon[ totalImages ];
for ( int i = 0; i < images.length; ++i )
images[ i ] = new ImageIcon( "images/deitel" + i + ".gif" );
startAnimation();
}
public void paintComponent( Graphics g )

{
super.paintComponent( g );
if ( images[ currentImage ].getImageLoadStatus() == MediaTracker.COMPLETE
) {
images[ currentImage ].paintIcon( this, g, 0, 0 );
currentImage = ( currentImage + 1 ) % totalImages;
}
}
public void actionPerformed( ActionEvent e )
{
repaint();
}
public void startAnimation()
{
if ( animationTimer == null ) {
currentImage = 0;
animationTimer = new Timer( animationDelay, this );
animationTimer.start();
}
else // continue from last image displayed
if ( ! animationTimer.isRunning() )
animationTimer.restart();
}
```

Conti.......

```
public void startAnimation()
{
if ( animationTimer == null ) {
currentImage = 0;
animationTimer = new Timer( animationDelay, this );
animationTimer.start();
}
else // continue from last image displayed
if ( ! animationTimer.isRunning() )
animationTimer.restart();
}

public void stopAnimation()
{
animationTimer.stop();
}

public Dimension getMinimumSize()
{
return getPreferredSize();
}

public Dimension getPreferredSize()
{
return new Dimension( 160, 80 );
  }

public static void main( String args[] )
{
LogoAnimator anim = new LogoAnimator();

JFrame app = new JFrame( "Animator test" );
app.getContentPane().add( anim, BorderLayout.CENTER );

app.addWindowListener(
new WindowAdapter() {
public void windowClosing( WindowEvent e )
{
System.exit( 0 );
}
}
);
```

25.4 Playing Audio Clips

```
// The constants 10 and 30 are used below to size the
// window 10 pixels wider than the animation and
// 30 pixels taller than the animation.
app.setSize( anim.getPreferredSize().width + 10,
anim.getPreferredSize().height + 30 );
}
}
```

Java can manipulate and play audio clips, it provides two mechanisms for playing sounds in an applet.

- Applets play method (plays the audio clip once only)
- The play method from the Audio Clip interface (can loop a clip)

```
import java.applet.*;
import java.awt.*;
import java.awt.event.*;
import javax.swing.*;
public class LoadAudioAndPlay extends JApplet {
private AudioClip sound1, sound2, currentSound;
private JButton playSound, loopSound, stopSound;
private JComboBox chooseSound;
// load the image when the applet begins executing
public void init()
{
Container c = getContentPane();
c.setLayout( new FlowLayout() );
String choices[] = { "Welcome", "Hi" };
chooseSound = new JComboBox( choices );
chooseSound.addItemListener(
new ItemListener() {
public void itemStateChanged( ItemEvent e )
{
currentSound.stop();
currentSound =
chooseSound.getSelectedIndex() == 0 ?
sound1 : sound2;
}
}
);
```

Conti.....

```
c.add( chooseSound );

ButtonHandler handler = new ButtonHandler();
playSound = new JButton( "Play" );
playSound.addActionListener( handler );
c.add( playSound );
loopSound = new JButton( "Loop" );
loopSound.addActionListener( handler );
c.add( loopSound );
stopSound = new JButton( "Stop" );
stopSound.addActionListener( handler );
c.add( stopSound );

sound1 = getAudioClip(
getDocumentBase(), "welcome.wav" );
sound2 = getAudioClip(
getDocumentBase(), "hi.au" );
currentSound = sound1;
}
// stop the sound when the user switches Web pages

// (i.e., be polite to the user)
public void stop()
{
currentSound.stop();
}
private class ButtonHandler implements ActionListener {
public void actionPerformed( ActionEvent e )
{
if ( e.getSource() == playSound )
currentSound.play();
else if ( e.getSource() == loopSound )
currentSound.loop();
else if ( e.getSource() == stopSound )
currentSound.stop();
}
}
}

## HTML Code
<html>
<applet code="LoadAudioAndPlay.class" width=350 height=50>
</applet>
</html>
```

25.5 JDBC

JDBC - Introduction

JDBC stands for Java Database Connectivity, which is a standard Java API for database-independent connectivity between the Java programming language and a wide range of databases.

The JDBC library includes APIs for each of the tasks mentioned below that are commonly associated with database usage.

- Making a connection to a database.
- Creating SQL or MySQL statements.
- Executing SQL or MySQL queries in the database.
- Viewing & Modifying the resulting records.

Fundamentally, JDBC is a specification that provides a complete set of interfaces that allows for portable access to an underlying database. Java can be used to write different types of executables, such as –

- Java Applications
- Java Applets
- Java Servlets
- Java ServerPages (JSPs)
- Enterprise JavaBeans (EJBs).

All of these different executables are able to use a JDBC driver to access a database, and take advantage of the stored data. JDBC provides the same capabilities as ODBC, allowing Java programs to contain database-independent code.

25.6 JDBC Architecture

The JDBC API supports both two-tier and three-tier processing models for database access but in general, JDBC Architecture consists of two layers

- **JDBC API:** This provides the application-to-JDBC Manager connection.
- **JDBC Driver API:** This supports the JDBC Manager-to-Driver Connection.

The JDBC API uses a driver manager and database-specific drivers to provide transparent connectivity to heterogeneous databases. The JDBC driver manager ensures that the correct driver is used to access each data source. The driver manager is capable of supporting multiple concurrent drivers connected to multiple heterogeneous databases.

Following is the architectural diagram, which shows the location of the driver manager with respect to the JDBC drivers and the Java application –

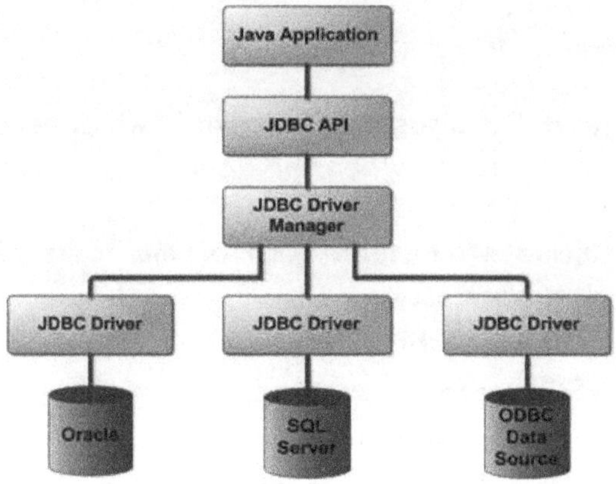

Fig: 25. 1

25.7 Common JDBC Components

The JDBC API provides the following interfaces and classes

- **DriverManager:** This class manages a list of database drivers. Matches connection requests from the java application with the proper database driver using communication sub protocol. The first driver that recognizes a certain sub protocol under JDBC will be used to establish a database Connection.
- **Driver:** This interface handles the communications with the database server. You will interact directly with Driver objects very rarely. Instead, you use DriverManager objects, which manages objects of this type. It also abstracts the details associated with working with Driver objects.
- **Connection:** This interface with all methods for contacting a database. The connection object represents communication context, i.e., all communication with database is through connection object only.
- **Statement:** You use objects created from this interface to submit the SQL statements to the database. Some derived interfaces accept parameters in addition to executing stored procedures.
- **ResultSet:** These objects hold data retrieved from a database after you execute an SQL query using Statement objects. It acts as an iterator to allow you to move through its data.
- **SQLException:** This class handles any errors that occur in a database application.

25.8 JDBC Driver

JDBC drivers implement the defined interfaces in the JDBC API, for interacting with your database server.

For example, using JDBC drivers enable you to open database connections and to interact with it by sending SQL or database commands then receiving results with Java.

The **Java.sql** package that ships with JDK, contains various classes with their behaviours defined and their actual implementations are done in third-party drivers. Third party vendors implement the **java.sql.Driver** interface in their database driver.

25.9 JDBC Drivers Types

JDBC driver implementations vary because of the wide variety of operating systems and hardware platforms in which Java operates. Sun has divided the implementation types into four categories, Types 1, 2, 3, and 4, which is explained below:

Type 1: JDBC-ODBC Bridge Driver

- In a Type 1 driver, a JDBC bridge is used to access ODBC drivers installed on each client machine. Using ODBC, requires configuring on your system a Data Source Name (DSN) that represents the target database.
- When Java first came out, this was a useful driver because most databases only supported ODBC access but now this type of driver is recommended only for experimental use or when no other alternative is available.

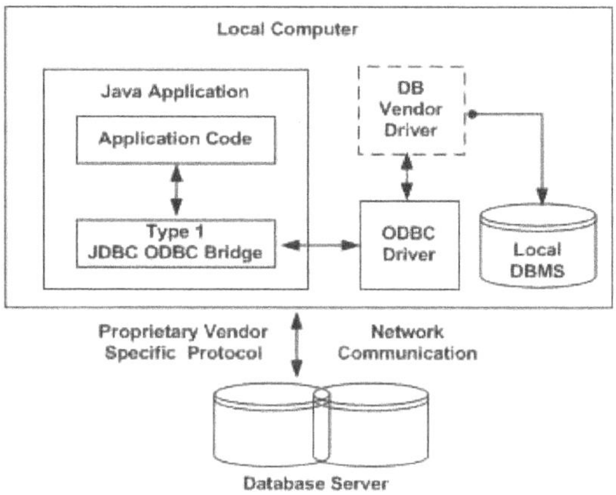

Fig: 25.2

Type 2: JDBC-Native API

- In a Type 2 driver, JDBC API calls are converted into native C/C++ API calls, which are unique to the database.
- These drivers are typically provided by the database vendors and used in the same manner as the JDBC-ODBC Bridge. The vendor-specific driver must be installed on each client machine.
- If we change the Database, we have to change the native API, as it is specific to a database and they are mostly obsolete now, but you may realize some speed increase with a Type 2 driver, because it eliminates ODBC's overhead.

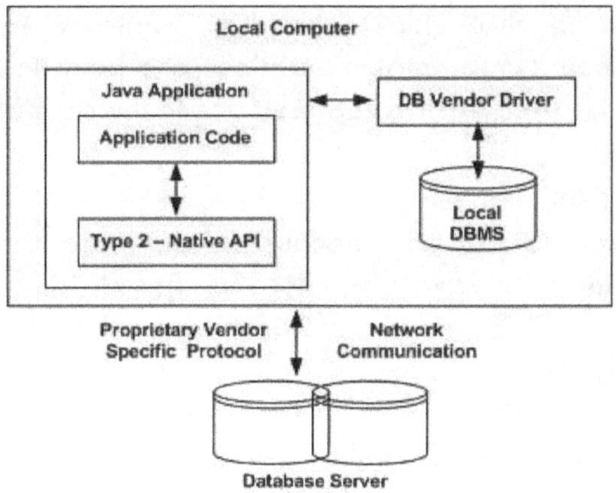

Fig:25.3

Type 3: JDBC-Net pure Java

- In a Type 3 driver, a three-tier approach is used to access databases. The JDBC clients use standard network sockets to communicate with a middleware application server.
- The socket information is then translated by the middleware application server into the call format required by the DBMS, and forwarded to the database server.
- This kind of driver is extremely flexible, since it requires no code installed on the client and a single driver can actually provide access to multiple databases.
- You can think of the application server as a JDBC "proxy," meaning that it makes calls for the client application. As a result, you need some knowledge of the application server's configuration in order to effectively use this driver type.

Your application server might use a Type 1, 2, or 4 drivers to communicate with the database, understanding the nuances will prove helpful.

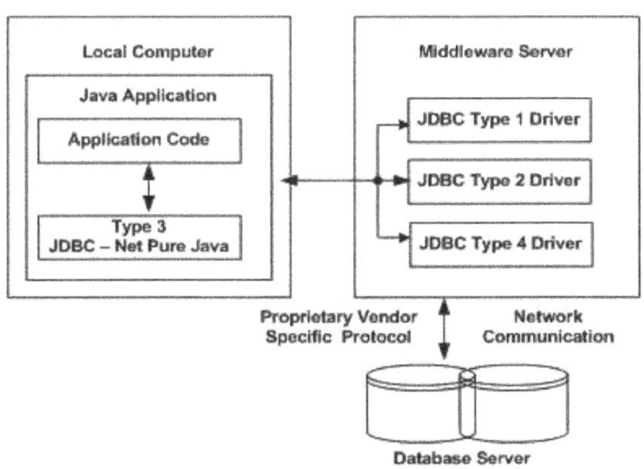

Fig: 25.4

Type 4: 100% Pure Java

- In a Type 4 driver, a pure Java-based driver communicates directly with the vendor's database through socket connection.
- This is the highest performance driver available for the database and is usually provided by the vendor itself.
- This kind of driver is extremely flexible; you don't need to install special software on the client or server. Further, these drivers can be downloaded dynamically.

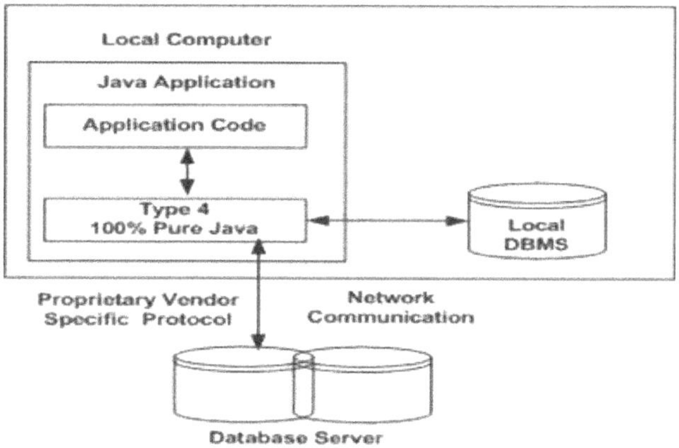

Fig: 25. 5

Which Driver should be used?

- If you are accessing one type of database, such as Oracle, Sybase, or IBM, the preferred driver type is 4.
- If your Java application is accessing multiple types of databases at the same time, type 3 is the preferred driver.
- Type 2 drivers are useful in situations, where a type 3 or type 4 driver is not available yet for your database.
- The type 1 driver is not considered a deployment-level driver, and is typically used for development and testing purposes only.

25.10 The Statement Objects

Creating Statement Object

Before you can use a Statement object to execute a SQL statement, you need to create one using the Connection object's createStatement( ) method, as in the following example

```
Statement stmt = null;
try {
stmt = conn.createStatement( );
. . .
}
catch (SQLException e) {
. . .
}
finally {
. . .
}
```

Once you've created a Statement object, you can then use it to execute an SQL statement with one of its three execute methods.

- **Boolean Execute (String SQL)**: Returns a boolean value of true if a ResultSet object can be retrieved; otherwise, it returns false. Use this method to execute SQL DDL statements or when you need to use truly dynamic SQL.
- **Int Executeupdate (String SQL)**: Returns the number of rows affected by the execution of the SQL statement. Use this method to execute SQL statements for which you expect to get a number of rows affected - for example, an INSERT, UPDATE, or DELETE statement.
- **ResultSet executeQuery (String SQL)**: Returns a ResultSet object. Use this method when you expect to get a result set, as you would with a SELECT statement.

Closing Statement Object

- Just as you close a Connection objects to save database resources, for the same reason you should also close the Statement object.
- A simple call to the close() method will do the job. If you close the Connection object first, it will close the Statement object as well. However, you should always explicitly close the Statement object to ensure proper cleanup.

```
Statement stmt = null;
try {
stmt = conn.createStatement( );
. . .
}
catch (SQLException e) {
. . .
}
finally {
stmt.close();
}
```

JDBC - Create Database Example

Sample Code

```
//STEP 1. Import required packages
import java.sql.*;

public class JDBCExample {
// JDBC driver name and database URL
static final String JDBC_DRIVER = "com.mysql.jdbc.Driver";
static final String DB_URL = "jdbc:mysql://localhost/";

// Database credentials
static final String USER = "username";
static final String PASS = "password";

public static void main(String[] args) {
Connection conn = null;
Statement stmt = null;
try{
//STEP 2: Register JDBC driver
Class.forName("com.mysql.jdbc.Driver");

//STEP 3: Open a connection
System.out.println("Connecting to database...");
conn = DriverManager.getConnection(DB_URL, USER, PASS);
try{
if(conn!=null)
conn.close();
}catch(SQLException se){
se.printStackTrace();
}//end finally try
}//end try
System.out.println("Goodbye!");
}//end main
}//end JDBCExample
```

CHAPTER- 25

JAVA SERVLETS

What is a Servlet?

Servlet can be described in many ways, depending on the context.

- Servlet is a technology i.e. used to create web application.
- Servlet is an API that provides many interfaces and classes including documentations.
- Servlet is an interface that must be implemented for creating any servlet.
- Servlet is a class that extend the capabilities of the servers and respond to the incoming request. It can respond to any type of requests.
- Servlet is a web component that is deployed on the server to create dynamic web page.

Advantages in comparison with the CGI.

- Performance is significantly better.
- Servlets execute within the address space of a Web server. It is not necessary to create a separate process to handle each client request.
- Servlets are platform-independent because they are written in Java.
- Java security manager on the server enforces a set of restrictions to protect the resources on a server machine. So servlets are trusted.
- The full functionality of the Java class libraries is available to a servlet. It can communicate with applets, databases, or other software via the sockets and RMI mechanisms that you have seen already.

26.1 Servlets Architecture

The following diagram shows the position of Servlets in a Web Application.

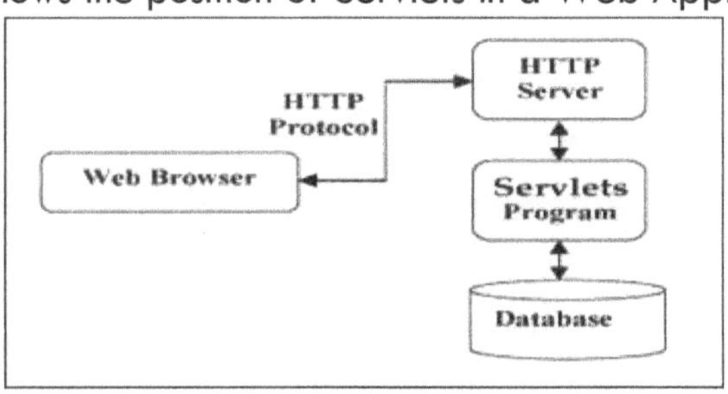

Fig: 26.1

26.2 Servlets Tasks

Servlets perform the following major tasks –

- Read the explicit data sent by the clients (browsers). This includes an HTML form on a Web page or it could also come from an applet or a custom HTTP client program.
- Read the implicit HTTP request data sent by the clients (browsers). This includes cookies, media types and compression schemes the browser understands, and so forth.
- Process the data and generate the results. This process may require talking to a database, executing an RMI or CORBA call, invoking a Web service, or computing the response directly.
- Send the explicit data (i.e., the document) to the clients (browsers). This document can be sent in a variety of formats, including text (HTML or XML), binary (GIF images), Excel, etc.
- Send the implicit HTTP response to the clients (browsers). This includes telling the browsers or other clients what type of document is being returned (e.g., HTML), setting cookies and caching parameters, and other such tasks.

26.3 Servlet Life Cycle

A servlet life cycle can be defined as the entire process from its creation till the destruction. The following are the paths followed by a servlet.

- The servlet is initialized by calling the **init()** method.
- The servlet calls **service()** method to process a client's request.
- The servlet is terminated by calling the **destroy()** method.
- Finally, servlet is garbage collected by the garbage collector of the JVM.

1) The init() Method

- The init method is called only once. It is called only when the servlet is created, and not called for any user requests afterwards. So, it is used for one-time initializations, just as with the init method of applets.
- The servlet is normally created when a user first invokes a URL corresponding to the servlet, but you can also specify that the servlet be loaded when the server is first started.
- When a user invokes a servlet, a single instance of each servlet gets created, with each user request resulting in a new thread that is handed off to doGet or doPost as appropriate. The init() method simply creates or loads some data that will be used throughout the life of the servlet.

The init method definition looks like this

```
public void init() throws ServletException {
// Initialization code...
}
```

2) The service() Method

- The service() method is the main method to perform the actual task. The servlet container (i.e. web server) calls the service() method to handle requests coming from the client(browsers) and to write the formatted response back to the client.
- Each time the server receives a request for a servlet, the server spawns a new thread and calls service. The service() method checks the HTTP request type **(GET, POST, PUT, DELETE,** etc.) and calls **doGet, doPost, doPut, doDelete,** etc. methods as appropriate.
- The service () method is called by the container and service method invokes doGe, doPost, doPut, doDelete, etc. methods as appropriate. So you have nothing to do with service() method but you override either doGet() or doPost() depending on what type of request you receive from the client.

Here is the signature of this method

```
public void service(ServletRequest request, ServletResponse response)
  throws ServletException, IOException {
}
```

The doGet() and doPost() are most frequently used methods within each service request. Here is the signature of these two methods.

i) The doGet() Method

A GET request results from a normal request for a URL or from an HTML form that has no METHOD specified and it should be handled by doGet() method.

```
public void doGet(HttpServletRequest request, HttpServletResponse response)
throws ServletException, IOException {
// Servlet code
}
```

ii) The doPost() Method

A POST request results from an HTML form that specifically lists POST as the METHOD and it should be handled by doPost() method.

```
public void doPost(HttpServletRequest request, HttpServletResponse response)
throws ServletException, IOException {
// Servlet code
}
```

3) The destroy() Method

- The destroy() method is called only once at the end of the life cycle of a servlet.
- This method gives your servlet a chance to close database connections, halt background threads, write cookie lists or hit counts to disk, and perform other such cleanup activities.

After the destroy() method is called, the servlet object is marked for garbage collection. The destroy method definition looks like this

```
public void destroy() {
// Finalization code...
}
```

Architecture Diagram

The following figure depicts a typical servlet life-cycle scenario.

- First the HTTP requests coming to the server are delegated to the servlet container.
- The servlet container loads the servlet before invoking the service() method.
- Then the servlet container handles multiple requests by spawning multiple threads, each thread executing the service() method of a single instance of the servlet.

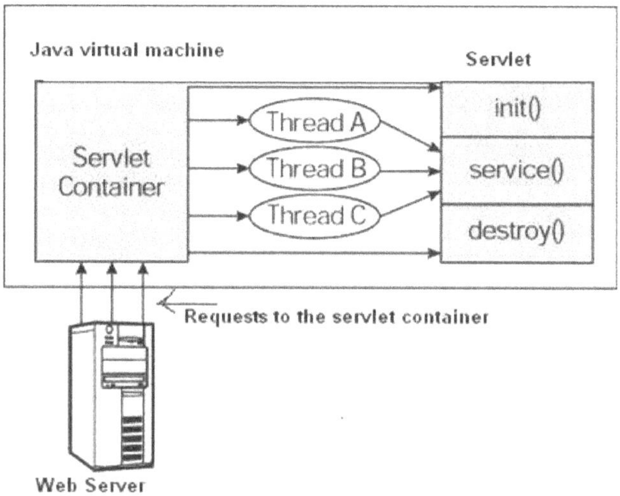

Fig: 26. 2

Servlets – Examples

Servlets are Java classes which service HTTP requests and implement the **javax.servlet.Servlet** interface. Web application developers typically write servlets that extend **javax.servlet.http.HttpServlet,** an abstract class that implements the Servlet interface and is specially designed to handle HTTP requests.

Sample Code

Following is the sample source code structure of a servlet example to show **"Hello World"**

```
// Import required java libraries
import java.io.*;
import javax.servlet.*;
import javax.servlet.http.*;

// Extend HttpServlet class
public class HelloWorld extends HttpServlet {

private String message;

public void init() throws ServletException {
// Do required initialization
message = "Hello World";
}
public void doGet(HttpServletRequest request, HttpServletResponse response)
throws ServletException, IOException {

// Set response content type
response.setContentType("text/html");
// Actual logic goes here.
PrintWriter out = response.getWriter();
out.println("<h1>" + message + "</h1>");
}
public void destroy() {
// do nothing.
}
}
```

26.4 Constituents of javax.servlet package

- The **javax.servlet** package is the core of the Servlet API.
- It contains the classes necessary for a standard, protocol-independent servlet. Every servlet must implement the Servlet interface in one form or another.
- The abstract GenericServlet class provides the framework for developing basic servlets. The package also includes a variety of utility classes that communicate with the server and the client

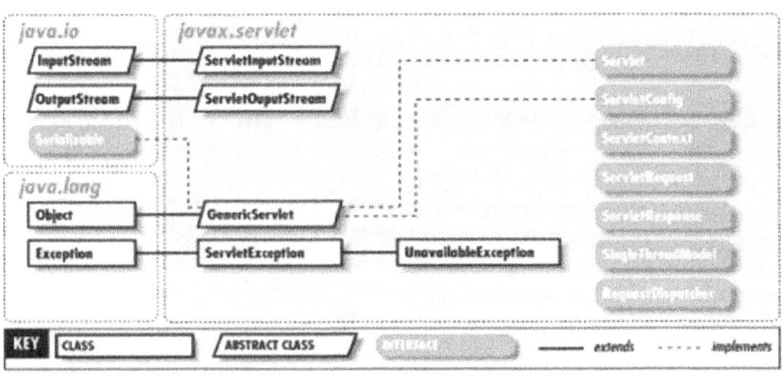

Fig: 26.3 The Javax.servlet Package

1) GenericServlet

The **GenericServlet** class provides a basic implementation of the Servlet and **ServletConfig** interfaces. If you are creating a protocol-independent servlet, you probably want to subclass this class rather than implement the Servlet interface directly. Note that the service() method is declared as abstract; this is the only method you have to override to implement a generic servlet.

```
public abstract class GenericServlet implements Servlet, ServletConfig, Serializable {
//        Public Constructors
public GenericServlet ();
//        Public Instance Methods
2.1       public void init () throws ServletException; empty
public void log (String msg);
2.1       public void log (String message, Throwable t);
//        Methods implementing Servlet
public void destroy ();
public ServletConfig getServletConfig ();
public String getServletInfo ();
public void init (ServletConfig config) throws ServletException;
public abstract void service (ServletRequest req, ServletResponse res) throws ServletException,
IOException;
//        Methods implementing ServletConfig
public String getInitParameter (String name);
public java.util.Enumeration getInitParameterNames ();
public ServletContext getServletContext ();
}
```

2) RequestDispatcher

RequestDispatcher allows a servlet to delegate some or all of the processing of a request to another resource on the web server. A RequestDispatcher object is retrieved by calling the getRequestDispatcher() method on the ServletContext object. The forward() method passes a request on to another servlet for processing, while the include() method includes the output of another servlet in the output of the current servlet.

```
public interface RequestDispatcher {
//        Public Instance Methods
public abstract void forward (ServletRequest request, ServletResponse response) throws
ServletException, IOException;
public abstract void include (ServletRequest request, ServletResponse response) throws
ServletException, IOException;
}
```

3) Servlet

The Servlet interface defines the basic structure of a **servlet.** All servlets implement this interface, either directly or by **subclassing** a class that does. The interface declares the basic servlet functionality: initializing a servlet, handling client requests, and destroying a servlet.

```
public interface Servlet {
//        Public Instance Methods
public abstract void destroy ();
public abstract ServletConfig getServletConfig ();
public abstract String getServletInfo ();
public abstract void init (ServletConfig config) throws ServletException;
public abstract void service (ServletRequest req, ServletResponse res) throws
ServletException, IOException;
}
```

4) ServletConfig

A **ServletConfig** object passes configuration information from the server to a **servlet.ServletConfig** supports initialization parameters (also known simply as init parameters) defined by the server administrator for a particular servlet.

```
public interface ServletConfig {
//        Public Instance Methods
public abstract String getInitParameter (String name);
public abstract java.util.Enumeration getInitParameterNames ();

public abstract ServletContext getServletContext ();
}
```

5) ServletContext

ServletContext defines methods that allow a servlet to interact with the host server. This includes reading server-specific attributes, finding information about particular files located on the server, and writing to the server log files. If there are several virtual servers running, each one may return a different ServletContext.

```
public interface ServletContext {
// Public Instance Methods
  public abstract Object getAttribute (String name);
  public abstract java.util.Enumeration getAttributeNames ();
  public abstract ServletContext getContext (String uripath);
  public abstract int getMajorVersion ();
  public abstract String getMimeType (String file);
  public abstract int getMinorVersion ();
  public abstract String getRealPath (String path);
  public abstract RequestDispatcher getRequestDispatcher (String urlpath);
  public abstract java.net.URL getResource (String path) throws
  java.net.MalformedURLException;
  public abstract java.io.InputStream getResourceAsStream (String path);
  public abstract String getServerInfo ();
  public abstract void log (String msg);
  public abstract void log (String message, Throwable throwable);
  public abstract void reaoveAttribute (String name);
  public abstract void setAttribute (String name, Object object);
// Deprecated Public Methods
# public abstract Servlet getServlet (String name) throws ServletException;
# public abstract java.util.Enumeration getServletNames ();
```

6) ServletException

A generic Exception class used for basic servlet errors. In version 2.1, a servlet can specify a Throwable root cause for this exception (using the constructors that accept Throwable parameters). The root cause can be retrieved with the getRootCause() method.

```
public class ServletException extends Exception {
//          Public Constructors
            public ServletException ();
            public ServletException (Throwable rootCause);
public ServletException (String message);
            public ServletException (String message, Throwable rootCause);

//          Public Instance Methods
            public Throwable getRootCause (); default:null
}
```

7) ServletInputStream

ServletInputStream provides an input stream for reading data from a client request. A servlet can get a ServletInputStream by calling the getInputStream() method of ServletRequest.

```
public abstract class ServletInputStream extends java.io.InputStream {
//          Protected Constructors
protected ServletInputStream ();
//          Public Instance Methods
public int readLine (byte ] b, int off, int len) throws IOException;
}
```

8) ServletOutputStream

ServletOutputStream provides an output stream for sending binary data back to a client. A servlet can get a ServletOutputStream by calling the getOutputStream() method of ServletResponse.

```
public abstract class
ServletOutputStream extends java.io.OutputStream {
// Protected Constructors
    protected ServletOutputStream ();
// Public Instance Methods
    public void print (long l) throws IOException;
    public void print (float f) throws IOException;
    public void print (double d) throws IOException;
    public void print (int i) throws IOException;
    public void print (String s) throws IOException;
    public void print (boolean b) throws IOException;
    public void print (char c) throws IOException;
    public void println (long l) throws IOException;
    public void println (float f) throws IOException;
    public void println (double d) throws IOException;
    public void println (int i) throws IOException;
    public void println (String s) throws IOException;
    public void println (boolean b) throws IOException;
```

9) ServletRequest

A **ServletRequest** object encapsulates information about a client request. The server passes a ServletRequest object to the service() method of a servlet. ServletRequest provides access to request parameters, such as form values or other request-specific parameters. These are accessed using the **getParameterNames(), getParameter(), and getParameterValues()** methods.

10) ServletResponse

The **ServletResponse** object sends MIME-encoded data back to the client. The interface defines a **getOutputStream()** method that returns a **ServletOutputStream** for sending binary data and a **getWriter()** method that returns a PrintWriter for sending textual data.

```
public interface ServletResponse {
//        Public Instance Methods
          public abstract String getCharacterEncoding ();
public abstract ServletOutputStream getOutputStream () throws IOException;

          public abstract PrintWriter getWriter () throws IOException;
public abstract void setContentLength (int len);
public abstract void setContentType (String type);
}
```

11) SingleThreadModel

SingleThreadModel is a tag interface that tells the server to create a pool of servlet instances to serve individual requests. In this case, the server ensures that each instance of the servlet handles only one service request at a time. **SingleThreadModel** provides easy thread safety, but imposes performance penalties.

```
public interface SingleThreadModel {
}
```

26.5 Cookies

Cookies are text files stored on the client computer and they are kept for various information tracking. Java Servlets transparently supports HTTP cookies.

There are three steps involved in identifying returning users –

- Server script sends a set of cookies to the browser. For example name, age, or identification number etc.
- Browser stores this information on local machine for future use.
- When next time browser sends any request to web server then it sends those cookies information to the server and server uses that information to identify the user.

26.6 Servlet Cookies Methods

Following is the list of useful methods which you can use while manipulating cookies in servlet.

S.No	Method & Description
1	**public void setDomain(String pattern)** This method sets the domain to which cookie applies.
2	**public String getDomain()** This method gets the domain to which cookie applies.
3	**public void setMaxAge(int expiry)** This method sets how much time (in seconds) should elapse before the cookie expires. If you don't set this, the cookie will last only for the current session.
4	**public int getMaxAge()** This method returns the maximum age of the cookie, specified in seconds, By default, -1 indicating the cookie will persist until browser shutdown.

5	**public String getName()** This method returns the name of the cookie. The name cannot be changed after creation.
6	**public void setValue(String newValue)** This method sets the value associated with the cookie
7	**public String getValue()** This method gets the value associated with the cookie.
8	**public void setPath(String uri)** This method sets the path to which this cookie applies. If you don't specify a path, the cookie is returned for all URLs in the same directory as the current page as well as all subdirectories.
9	**public String getPath()** This method gets the path to which this cookie applies.
10	**public void setSecure(boolean flag)** This method sets the boolean value indicating whether the cookie should only be sent over encrypted (i.e. SSL) connections.
11	**public void setComment(String purpose)** This method specifies a comment that describes a cookie's purpose. The comment is useful if the browser presents the cookie to the user.
12	**public String getComment()** This method returns the comment describing the purpose of this cookie, or null if the cookie has no comment.

Table: 26.1

26.7 Setting Cookies with Servlet

Setting cookies with Servlet involves three steps:

(1) Creating a Cookie object – you call the Cookie constructor with a cookie name and a cookie value, both of which are strings.

```
Cookie cookie = new Cookie("key","value");
```

Keep in mind, neither the name nor the value should contain white space or any of the following characters:

```
[] () = , " / ? @ : ;
```

(2) Setting the maximum age – you use **setMaxAge** to specify how long (in seconds) the cookie should be valid. Following would set up a cookie for 24 hours.

```
cookie.setMaxAge(60 * 60 * 24);
```

(3) Sending the Cookie into the HTTP response headers – you use response.addCookie to add cookies in the HTTP response header as follows:

```
response.addCookie(cookie);
```

Example

```
// Import required java libraries
import java.io.*;
import javax.servlet.*;
import javax.servlet.http.*;
// Extend HttpServlet class
public class HelloForm extends HttpServlet {
public void doGet(HttpServletRequest request, HttpServletResponse response)
throws ServletException, IOException {
// Create cookies for first and last names.
Cookie firstName = new Cookie("first_name", request.getParameter("first_name"));
Cookie lastName = new Cookie("last_name", request.getParameter("last_name"));
// Set expiry date after 24 Hrs for both the cookies.
firstName.setMaxAge(60*60*24);
lastName.setMaxAge(60*60*24);
// Add both the cookies in the response header.
response.addCookie( firstName );
response.addCookie( lastName );
// Set response content type
response.setContentType("text/html");
PrintWriter out = response.getWriter();
String title = "Setting Cookies Example";
String docType =
"<!doctype html public \"-//w3c//dtd html 4.0 " + "transitional//en\">\n";

out.println(docType +

"<html>\n" +
"<head>
<title>" + title + "</title>
</head>\n" +
```

Conti......

```
"<body bgcolor = \"#f0f0f0\">\n" +
"<h1 align = \"center\">" + title + "</h1>\n" +
"<ul>\n" +
"  <li><b>First Name</b>: "
+ request.getParameter("first_name") + "\n" +
"  <li><b>Last Name</b>: "
+ request.getParameter("last_name") + "\n" +
"</ul>\n" +
"</body>
 </html>"
);
}
}
```

Compile the above servlet **HelloForm** and create appropriate entry in web.xml file and finally try following **HTML page** to call servlet.

```
<html>
 <body>
<form action = "HelloForm" method = "GET">
First Name: <input type = "text" name = "first_name">
<br />
Last Name: <input type = "text" name = "last_name" />
<input type = "submit" value = "Submit" />
</form>
</body>
</html>
```

26.8 Session Tracking

A webserver can assign a unique session ID as a cookie to each web client and for subsequent requests from the client they can be recognized using the recieved cookie.

This may not be an effective way because many time browsers do not support a cookie, so I would not recommend using this procedure to maintain the sessions.

Hidden Form Fields

A web server can send a hidden HTML form field along with a unique session ID as follows:

```
<input type = "hidden" name = "sessionid" value = "12345">
```

This entry means that, when the form is submitted, the specified name and value are automatically included in the GET or POST data. Each time when web browser sends request back, then session_id value can be used to keep the track of different web browsers.

This could be an effective way of keeping track of the session but clicking on a regular (<A HREF...>) hypertext link does not result in a form submission, so hidden form fields also cannot support general session tracking.

26.9 URL Rewriting

You can append some extra data on the end of each URL that identifies the session, and the server can associate that session identifier with data it has stored about that session.

URL rewriting is a better way to maintain sessions and it works even when browsers don't support cookies. The drawback of URL re-writing is that you would have to generate every URL dynamically to assign a session ID, even in case of a simple static HTML page.

26.10 The HttpSession Object

Apart from the above mentioned three ways, servlet provides HttpSession Interface which provides a way to identify a user across more than one page request or visit to a Web site and to store information about that user.

The servlet container uses this interface to create a session between an HTTP client and an HTTP server. The session persists for a specified time period, across more than one connection or page request from the user. You would get HttpSession object by calling the public method getSession() of HttpServletRequest, as below

```
HttpSession session = request.getSession();
```

You need to call request.getSession() before you send any document content to the client. Here is a summary of the important methods available through HttpSession object

S.No	Method & Description
1	**public Object getAttribute(String name)** This method returns the object bound with the specified name in this session or null if no object is bound under the name.
2	**public Enumeration getAttributeNames()** This method returns Enumeration of String objects containing the names of all the objects bound to this session.
3	**public long getCreationTime()** This method returns the time when this session was created, measured in milliseconds since midnight January 1, 1970 GMT.
4	**public String getId()** This method returns a string containing the unique identifier assigned to this session.
5	**public long getLastAccessedTime()** This method returns the last accessed time of the session, in the format of milliseconds since midnight January 1, 1970 GMT
6	**public int getMaxInactiveInterval()** This method returns the maximum time interval (seconds), that the servlet container will keep the session open between client accesses.
7	**public void invalidate()** This method invalidates this session and unbinds any objects bound to it.
8	**public boolean isNew(** This method returns true if the client does not yet know about the session or if the client chooses not to join the session.
9	**public void removeAttribute(String name)** This method removes the object bound with the specified name from this session.
10	**public void setAttribute(String name, Object value)** This method binds an object to this session, using the name specified.
11	**public void setMaxInactiveInterval(int interval)** This method specifies the time, in seconds, between client requests before the servlet container will invalidate this session.

Table: 26. 2

Session Tracking Example

This example describes how to use the **HttpSession** object to find out the creation time and the last-accessed time for a session. We would associate a new session with the request if one does not already exist.

```
// Import required java libraries
import java.io.*;
import javax.servlet.*;
import javax.servlet.http.*;
import java.util.*;

// Extend HttpServlet class
public class SessionTrack extends HttpServlet {
public void doGet(HttpServletRequest request, HttpServletResponse response)
throws ServletException, IOException {
// Create a session object if it is already not  created.
HttpSession session = request.getSession(true);
// Get session creation time.
Date createTime = new Date(session.getCreationTime());
// Get last access time of this web page.
Date lastAccessTime = new Date(session.getLastAccessedTime());
String title = "Welcome Back to my website";
Integer visitCount = new Integer(0);
String visitCountKey = new String("visitCount");
String userIDKey = new String("userID");
String userID = new String("ABCD");
// Check if this is new comer on your web page.
if (session.isNew()) {
title = "Welcome to my website";
session.setAttribute(userIDKey, userID);
} else {
```

```
visitCount = (Integer)session.getAttribute(visitCountKey);
visitCount = visitCount + 1;
userID = (String)session.getAttribute(userIDKey);
}
session.setAttribute(visitCountKey,  visitCount);
// Set response content type
response.setContentType("text/html");
PrintWriter out = response.getWriter();
String docType =
"<!doctype html public \"-//w3c//dtd html 4.0 " +
"transitional//en\">\n";
out.println(docType +
"<html>\n" +
"<head><title>" + title + "</title></head>\n" +
"<body bgcolor = \"#f0f0f0\">\n" +
"<h1 align = \"center\">" + title + "</h1>\n" +
"<h2 align = \"center\">Session Infomation</h2>\n" +
"<table border = \"1\" align = \"center\">\n" +
"<tr bgcolor = \"#949494\">\n" +
"  <th>Session info</th><th>value</th>
</tr>\n" +

"<tr>\n" +
"  <td>id</td>\n" +
"  <td>" + session.getId() + "</td>
</tr>\n" +
"<tr>\n" +
"  <td>Creation Time</td>\n" +
"  <td>" + createTime + "  </td>
</tr>\n" +
```

Conti....

```
"<tr>\n" +
"  <td>Time of Last Access</td>\n" +
"  <td>" + lastAccessTime + "  </td>
</tr>\n" +
"<tr>\n" +
"  <td>User ID</td>\n" +
"  <td>" + userID + "  </td>
</tr>\n" +

"<tr>\n" +
"  <td>Number of visits</td>\n" +
"  <td>" + visitCount + "</td>
</tr>\n" +
"</table>\n" +
"</body>
</html>"
);
}
}
```

CHAPTER-26

JAVA SERVER PAGES

What is JSP?

- JavaServer Pages (JSP) is a technology for developing Webpages that supports dynamic content. This helps developers insert java code in HTML pages by making use of special JSP tags, most of which start with <% **and end with %>.**
- A JavaServer Pages component is a type of Java servlet that is designed to fulfill the role of a user interface for a Java web application. Web developers write JSPs as text files that combine HTML or XHTML code, XML elements, and embedded JSP actions and commands.
- Using JSP, you can collect input from users through Webpage forms, present records from a database or another source, and create Webpages dynamically.
- JSP tags can be used for a variety of purposes, such as retrieving information from a database or registering user preferences, accessing JavaBeans components, passing control between pages, and sharing information between requests, pages etc.

Why Use JSP?

- JavaServer Pages often serve the same purpose as programs implemented using the Common Gateway Interface (CGI). But JSP offers several advantages in comparison with the CGI.
- Performance is significantly better because JSP allows embedding Dynamic Elements in HTML Pages itself instead of having separate CGI files.
- JSP are always compiled before they are processed by the server unlike CGI/Perl which requires the server to load an interpreter and the target script each time the page is requested.
- JavaServer Pages are built on top of the Java Servlets API, so like Servlets, JSP also has access to all the powerful Enterprise Java APIs, including **JDBC, JNDI, EJB, JAXP,** etc.
- JSP pages can be used in combination with servlets that handle the business logic, the model supported by Java servlet template engines.

Advantages of JSP

1) vs. Active Server Pages (ASP)

The advantages of JSP are twofold. First, the dynamic part is written in Java, not Visual Basic or other MS specific language, so it is more powerful and easier to use. Second, it is portable to other operating systems and non-Microsoft Web servers.

2) vs. Pure Servlets

It is more convenient to write (and to modify!) regular HTML than to have plenty of println statements that generate the HTML.

3) vs. Server-Side Includes (SSI)

SSI is really only intended for simple inclusions, not for "real" programs that use form data, make database connections.

4) vs. Static HTML

Regular HTML, of course, cannot contain dynamic information.

5) vs. JavaScript

JavaScript can generate HTML dynamically on the client but can hardly interact with the web server to perform complex tasks like database access and image processing etc.

27.1 JSP – Architecture

A JSP container works with the Web server to provide the runtime environment and other services a JSP needs. It knows how to understand the special elements that are part of JSPs.

Following diagram shows the position of JSP container and JSP files in a Web application.

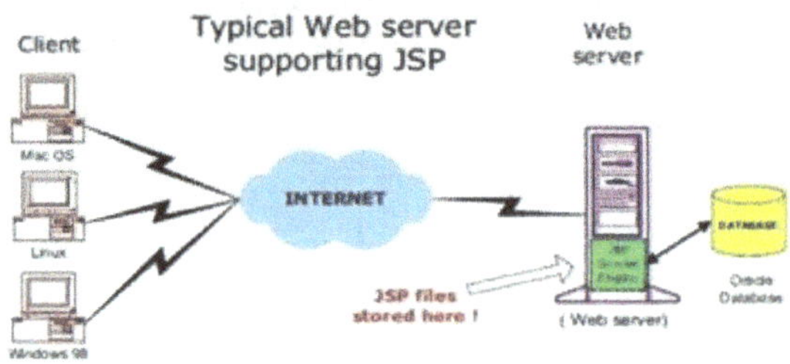

Fig: 27. 1

27.2 JSP Processing

- As with a normal page, your browser sends an HTTP request to the web server. The web server recognizes that the HTTP request is for a JSP page and forwards it to a JSP engine. This is done by using the URL or JSP page which ends with **.jsp** instead of **.html.**
- The JSP engine loads the JSP page from disk and converts it into servlet content. This conversion is very simple in which all template text is converted to **println( )** statements and all JSP elements are converted to Java code. This code implements the corresponding dynamic behavior of the page.
- The JSP engine compiles the servlet into an executable class and forwards the original request to a servlet engine.
- A part of the web server called the servlet engine loads the Servlet class and executes it. During execution, the servlet produces an output in HTML format. The output is furthur passed on to the web server by the servlet engine inside an HTTP response.

- The web server forwards the HTTP response to your browser in terms of static HTML content.
- Finally, the web browser handles the dynamically-generated HTML page inside the HTTP response exactly as if it were a static page.

All the above mentioned steps can be seen in the following diagram

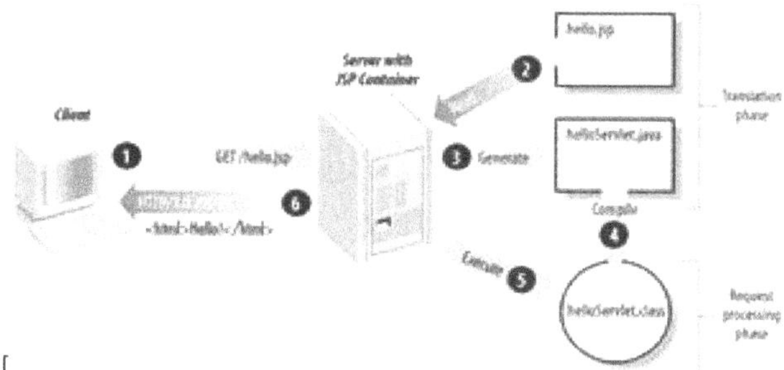

[

Fig: 27.2

The JSP engine checks to see whether a servlet for a JSP file already exists and whether the modification date on the JSP is older than the servlet. If the JSP is older than its generated servlet, the JSP container assumes that the JSP hasn't changed and that the generated servlet still matches the JSP's contents. This makes the process more efficient than with the other scripting languages (such as PHP) and therefore faster.

27.3 JSP – Lifecycle

The four major phases of a JSP life cycle are very similar to the Servlet Life Cycle. The four phases have been described below,

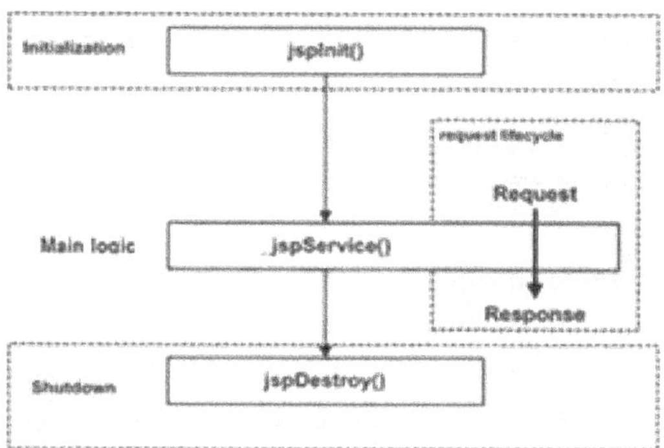

Fig: 27. 3

The following are the paths followed by a JSP –

- Compilation
- Initialization
- Execution
- Cleanup

JSP Compilation

When a browser asks for a JSP, the JSP engine first checks to see whether it needs to compile the page. If the page has never been compiled, or if the JSP has been modified since it was last compiled, the JSP engine compiles the page.

The compilation process involves three steps –

- Parsing the JSP.
- Turning the JSP into a servlet.
- Compiling the servlet.

JSP Initialization

When a container loads a JSP it invokes the **jspInit()** method before servicing any requests. If you need to perform JSP-specific initialization, override the **jspInit()** method

```
public void jspInit(){
// Initialization code...
}
```

JSP Execution

- This phase of the JSP life cycle represents all interactions with requests until the JSP is destroyed.
- Whenever a browser requests a JSP and the page has been loaded and initialized, the JSP engine invokes the _jspService() method in the JSP.
- The **jspService()** method takes an **HttpServletRequest** and an **HttpServletResponse** as its parameters as follows –

```
void _jspService(HttpServletRequest request, HttpServletResponse response) {
// Service handling code...
}
```

The _jspService() method of a JSP is invoked on request basis. This is responsible for generating the response for that request and this method is also responsible for generating responses to all seven of the HTTP methods, i.e, **GET, POST, DELETE,** etc.

JSP Cleanup

- The destruction phase of the JSP life cycle represents when a JSP is being removed from use by a container.
- The **jspDestroy()** method is the JSP equivalent of the destroy method for servlets. Override jspDestroy when you need to perform any cleanup, such as releasing database connections or closing open files.

The jspDestroy() method has the following form –

```
public void jspDestroy() {
// Your cleanup code goes here.
}
```

27.4 JSP – Syntax

Elements of JSP

1) The Scriptlet

A scriptlet can contain any number of JAVA language statements, variable or method declarations, or expressions that are valid in the page scripting language.

Following is the syntax of Scriptlet,

```
<% code fragment %>
```

You can write the XML equivalent of the above syntax as follows,

```
<jsp:scriptlet>
code fragment
</jsp:scriptlet>
```

Any text, HTML tags, or JSP elements you write must be outside the scriptlet. Following is the simple and first example for JSP,

```
<html>
<head><title>Hello World</title></head>

<body>
Hello World!<br/>
<%
out.println("Your IP address is " + request.getRemoteAddr());
%>
</body>
</html>
```

2) JSP Declarations

A declaration declares one or more variables or methods that you can use in Java code later in the JSP file. You must declare the variable or method before you use it in the JSP file.

Following is the syntax for JSP Declarations

```
<%! declaration; [ declaration; ]+ ... %>
```

You can write the XML equivalent of the above syntax as follows

```
<jsp:declaration>
  code fragment
</jsp:declaration>
```

Following is an example for JSP Declarations

```
<%! int i = 0; %>

<%! int a, b, c; %>

<%! Circle a = new Circle(2.0); %>
```

3) JSP Expression

A JSP expression element contains a scripting language expression that is evaluated, converted to a String, and inserted where the expression appears in the JSP file.

Because the value of an expression is converted to a String, you can use an expression within a line of text, whether or not it is tagged with HTML, in a JSP file.

The expression element can contain any expression that is valid according to the Java Language Specification but you cannot use a semicolon to end an expression.

Following is the syntax of JSP Expression

```
<%= expression %>
```

You can write the XML equivalent of the above syntax as follows

```
<jsp:expression>
  expression
</jsp:expression>
```

Following example shows a JSP Expression

```
<html>
  <head><title>A Comment Test</title></head>

  <body>
    <p>Today's date: <%= (new java.util.Date()).toLocaleString()%></p>
  </body>
</html>
```

4) JSP Comments

JSP comment marks text or statements that the JSP container should ignore. A JSP comment is useful when you want to hide or "comment out", a part of your JSP page.

Following is the syntax of the JSP comments

```
<%-- This is JSP comment --%>
```

Following example shows the JSP Comments

```
<html>
  <head><title>A Comment Test</title></head>

  <body>
    <h2>A Test of Comments</h2>
    <%-- This comment will not be visible in the page source --%>
  </body>
</html>
```

27.5 JSP - Implicit Objects

These Objects are the Java objects that the JSP Container makes available to the developers in each page and the developer can call them directly without being explicitly declared. JSP Implicit Objects are also called **pre-defined variables.**

Following table lists out the nine Implicit Objects that JSP supports:

S.No	Object & Description
1	**Request** This is the **HttpServletRequest** object associated with the request.
2	**Response** This is the **HttpServletResponse** object associated with the response to the client.
3	**Out** This is the **PrintWriter** object used to send output to the client.
4	**Session** This is the **HttpSession** object associated with the request.
5	**Application** This is the **ServletContext** object associated with the application context.
6	**Config** This is the **ServletConfig** object associated with the page.
7	**pageContext** This encapsulates use of server-specific features like higher performance **JspWriters.**
8	**Page** This is simply a synonym for **this,** and is used to call the methods defined by the translated servlet class.
9	**Exception** The **Exception object** allows the exception data to be accessed by designated JSP.

Table: 27.1

The request Object

- The request object is an instance of a **javax.servlet.http.HttpServletRequest** object. Each time a client requests a page the JSP engine creates a new object to represent that request.

- The request object provides methods to get the HTTP header information including form data, cookies, HTTP methods etc.

The response Object

- The response object is an instance of a **javax.servlet.http.HttpServletResponse** object. Just as the server creates the request object, it also creates an object to represent the response to the client.
- The response object also defines the interfaces that deal with creating new HTTP headers. Through this object the JSP programmer can add new cookies or date stamps, HTTP status codes, etc.

The out Object

- The out implicit object is an instance of a **javax.servlet.jsp.JspWriter** object and is used to send content in a response.
- The initial JspWriter object is instantiated differently depending on whether the page is buffered or not. Buffering can be easily turned off by using the **buffered = 'false'** attribute of the page directive.
- The JspWriter object contains most of the same methods as the **java.io.PrintWriter** class. However, JspWriter has some additional methods designed to deal with buffering. Unlike the PrintWriter object, JspWriter throws **IOExceptions.**
- Following table lists out the important methods that we will use to write **boolean char, int, double, object, String,** etc.

27.6 JSP Scripting

JSP Scripting elements are written inside <% %> tags. These code inside <% %> tags are processed by the JSP engine during translation of the JSP page. Any other text in the JSP page is considered as HTML code or plain text.

Example:

```
<html>
 <head>
 <title>My First JSP Page</title>
 </head>
  <%
 int count = 0;
 %>
 <body>
 Page Count is  <% out.println(++count); %>
 </body>
```

Just to experiment, try removing the <% %> scriplet tag from the above code and run it as JSP. You will see that everything is printed as it is on the browser, because without the scriplet tag, everything is considered plain HTML code.

There are five different types of scripting elements:

Scripting Element	Example
Comment	<%-- comment --%>
Directive	<%@ directive %>
Declaration	<%! declarations %>
Scriptlet	<% scriplets %>
Expression	<%= expression %>

Table: 27. 2

27.7 Standard Actions

These actions use constructs in XML syntax to control the behavior of the servlet engine. You can dynamically insert a file, reuse JavaBeans components, forward the user to another page, or generate HTML for the Java plugin. There is only one syntax for the Action element, as it conforms to the XML standard,

```
<jsp:action_name attribute = "value" />
```

Action elements are basically predefined functions. The following table lists out the available JSP actions,

S.No	Syntax & Purpose
1	**jsp:include** Includes a file at the time the page is requested.
2	**jsp:useBean** Finds or instantiates a JavaBean.
3	**jsp:setProperty** Sets the property of a JavaBean.
4	**jsp:getProperty** Inserts the property of a JavaBean into the output.
5	**jsp:forward** Forwards the requester to a new page.
6	**jsp:plugin** Generates browser-specific code that makes an OBJECT or EMBED tag for the Java plugin.
7	**jsp:element** Defines XML elements dynamically.
8	**jsp:attribute** Defines dynamically-defined XML element's attribute.
9	**jsp:body** Defines dynamically-defined XML element's body.
10	**jsp:text** Used to write template text in JSP pages and documents.

Table: 27. 3

27.8 Common Attributes

There are two attributes that are common to all Action elements: the id attribute and the scope attribute.

1) Id attribute

The id attribute uniquely identifies the Action element, and allows the action to be referenced inside the JSP page. If the Action creates an instance of an object, the id value can be used to reference it through the implicit object **PageContext.**

2) Scope attribute

This attribute identifies the lifecycle of the Action element. The id attribute and the scope attribute are directly related, as the scope attribute determines the lifespan of the object associated with the id. The scope attribute has four possible values: **(a) page, (b) request, (c)session, and (d) application.**

The <jsp:include> Action:

This action lets you insert files into the page being generated. The syntax looks like this,

```
<jsp:include page = "relative URL" flush = "true" />
```

Unlike the include directive, which inserts the file at the time the JSP page is translated into a servlet, this action inserts the file at the time the page is requested. Following table lists out the attributes associated with the include action

S.No	Attribute & Description
1	**Page** The relative URL of the page to be included.
2	**Flush** The boolean attribute determines whether the included resource has its buffer flushed before it is included.

Table: 27. 4

Example

Let us define the following two files **(a) date.jsp and (b) main.jsp** as follows
Following is the content of the **date.jsp** file

```
<p>Today's date: <%= (new java.util.Date()).toLocaleString()%></p>
```

Following is the content of the **main.jsp** file

```
<html>
  <head>
    <title>The include Action Example</title>
  </head>

  <body>
    <center>
      <h2>The include action Example</h2>
      <jsp:include page = "date.jsp" flush = "true" />
    </center>
  </body>
</html>
```

The <jsp:useBean> Action

The **useBean** action is quite versatile. It first searches for an existing object utilizing the id and scope variables. If an object is not found, it then tries to create the specified object. The simplest way to load a bean is as follows

```
<jsp:useBean id = "name" class = "package.class" />
```

Once a bean class is loaded, you can use **jsp:setProperty** and **jsp:getProperty** actions to modify and retrieve the bean properties.

Following table lists out the attributes associated with the **useBean** action

S.No	Attribute & Description
1	**Class** Designates the full package name of the bean.
2	**Type** Specifies the type of the variable that will refer to the object.
3	**beanName** Gives the name of the bean as specified by the instantiate () method of the **java.beans.Beans** class.

Table: 27. 5

The <jsp:setProperty> Action

The **setProperty** action sets the properties of a Bean. The Bean must have been previously defined before this action. There are two basic ways to use the setProperty action:

```
<jsp:useBean id = "myName" ... />
...
<jsp:setProperty name = "myName" property = "someProperty" .../>
```

You can use **jsp:setProperty** after, but outside of a **jsp:useBean** element, as given below:

In this case, the **jsp:setProperty** is executed regardless of whether a new bean was instantiated or an existing bean was found.

A second context in which jsp:setProperty can appear is inside the body of a **jsp:useBean**element, as given below

```
<jsp:useBean id = "myName" ... >

  ...

  <jsp:setProperty name = "myName" property = "someProperty" .../>

</jsp:useBean>
```

Here, the **jsp:setProperty** is executed only if a new object was instantiated, not if an existing one was found.

Following table lists out the attributes associated with the **setProperty** action

S.No	Attribute & Description
1	**Name** Designates the bean the property of which will be set. The Bean must have been previously defined.
2	**Property** Indicates the property you want to set. A value of "*" means that all request parameters whose names match bean property names will be passed to the appropriate setter methods.
3	**Value** The value that is to be assigned to the given property. The the parameter's value is null, or the parameter does not exist, the setProperty action is ignored.
4	**Param** The param attribute is the name of the request parameter whose value the property is to receive. You can't use both value and param, but it is permissible to use neither.

Table: 27. 6

The <jsp:getProperty> Action

The getProperty action is used to retrieve the value of a given property and converts it to a string, and finally inserts it into the output.

The getProperty action has only two attributes, both of which are required. The syntax of the getProperty action is as follows

```
<jsp:useBean id = "myName" ... />

...

<jsp:getProperty name = "myName" property = "someProperty" .../>
```

Following table lists out the required attributes associated with the **getProperty** action

S.No	Attribute & Description
1	**name** The name of the Bean that has a property to be retrieved. The Bean must have been previously defined.
2	**property** The property attribute is the name of the Bean property to be retrieved.

Table: 27. 7

The <jsp:forward> Action

The **forward** action terminates the action of the current page and forwards the request to another resource such as a static page, another JSP page, or a Java Servlet. Following is the syntax of the **forward** action

```
<jsp:forward page = "Relative URL" />
```

Following table lists out the required attributes associated with the forward action:

S.No	Attribute & Description
1	**Page** Should consist of a relative URL of another resource such as a static page, another JSP page, or a Java Servlet.

Table: 27.8

27.9 JSP-Directives

These directives provide directions and instructions to the container, telling it how to handle certain aspects of the JSP processing.

A JSP directive affects the overall structure of the servlet class. It usually has the following form:

```
<%@ directive attribute = "value" %>
```

Directives can have a number of attributes which you can list down as key-value pairs and separated by commas.

The blanks between the @ symbol and the directive name, and between the last attribute and the closing %>, are optional.

There are three types of directive tag:

S.No	Directive & Description
1	**<%@ page ... %>** Defines page-dependent attributes, such as scripting language, error page, and buffering requirements.
2	**<%@ include ... %>** Includes a file during the translation phase.
3	**<%@ taglib ... %>** Declares a tag library, containing custom actions, used in the page

Table: 27. 9

JSP - the page Directive

The **page** directive is used to provide instructions to the container. These instructions pertain to the current JSP page. You may code page directives anywhere in your JSP page. By convention, page directives are coded at the top of the JSP page.

Following is the basic syntax of the page directive:

```
<%@ page attribute = "value" %>
```

You can write the XML equivalent of the above syntax as follows

```
<jsp:directive.page attribute = "value" />
```

Attributes

Following table lists out the attributes associated with the **page directive**

S.No	Attribute & Purpose
1	**Buffer** Specifies a buffering model for the output stream.
2	**autoFlush** Controls the behavior of the servlet output buffer.
3	**contentType** Defines the character encoding scheme.
4	**errorPage** Defines the URL of another JSP that reports on Java unchecked runtime exceptions.
5	**isErrorPage** Indicates if this JSP page is a URL specified by another JSP page's errorPage attribute.
6	**Extends** Specifies a superclass that the generated servlet must extend.
7	**Import** Specifies a list of packages or classes for use in the JSP as the Java import statement does for Java classes.
8	**Info** Defines a string that can be accessed with the servlet's getServletInfo() method.
9	**isThreadSafe** Defines the threading model for the generated servlet.
10	**Language** Defines the programming language used in the JSP page.
11	**Session** Specifies whether or not the JSP page participates in HTTP sessions
12	**isELIgnored** Specifies whether or not the EL expression within the JSP page will be ignored.
13	**isScriptingEnabled** Determines if the scripting elements are allowed for use.

Table: 27.10

The include Directive

The include directive is used to include a file during the translation phase. This directive tells the container to merge the content of other external files with the current JSP during the translation phase. The general usage form of this directive is as follows:

```
<%@ include file = "relative url" >
```

The filename in the include directive is actually a relative URL. If you just specify a filename with no associated path, the JSP compiler assumes that the file is in the same directory as your JSP.

You can write the XML equivalent of the above syntax as follows:

```
<jsp:directive.include file = "relative url" />
```

The taglib Directive

The **JavaServer** Pages API allow you to define custom JSP tags that look like HTML or XML tags and a tag library is a set of user-defined tags that implement custom behavior.

The **taglib** directive declares that your JSP page uses a set of custom tags, identifies the location of the library, and provides means for identifying the custom tags in your JSP page.

The taglib directive follows the syntax given below:

```
<%@ taglib uri="uri" prefix = "prefixOfTag" >
```

Here, the **uri** attribute value resolves to a location the container understands and the **prefix** attribute informs a container what bits of markup are custom actions.

You can write the XML equivalent of the above syntax as follows:

```
<jsp:directive.taglib uri = "uri" prefix = "prefixOfTag" />
```

CHAPTER- 27

REMOTE METHOD INVOCATION

What is RMI?

RMI stands for **Remote Method Invocation.** It is a mechanism that allows an object residing in one system (JVM) to access/invoke an object running on another JVM.

RMI is used to build distributed applications; it provides remote communication between Java programs. It is provided in the package **java.rmi.**

28.1 Architecture of an RMI Application

In an RMI application, we write two programs, a **server program** (resides on the server) and a **client program** (resides on the client).

- Inside the server program, a remote object is created and reference of that object is made available for the client (using the registry).
- The client program requests the remote objects on the server and tries to invoke its methods.

The following diagram shows the architecture of an RMI application:

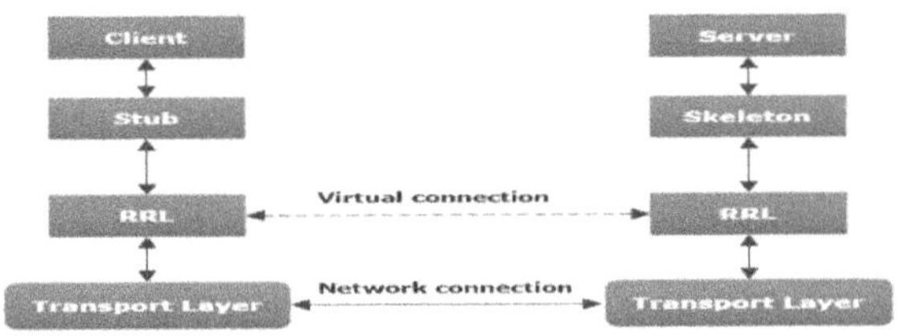

Fig: 28. 1

Let us now discuss the components of this architecture.

- **Transport Layer** –This layer connects the client and the server. It manages the existing connection and also sets up new connections.
- **Stub** – A stub is a representation (proxy) of the remote object at client. It resides in the client system; it acts as a gateway for the client program.
- **Skeleton** –This is the object which resides on the server side. **Stub** communicates with this skeleton to pass request to the remote object.
- **RRL (Remote Reference Layer)** – It is the layer which manages the references made by the client to the remote object.

28.2 Working of an RMI Application

The following points summarize how an RMI application works:

- When the client makes a call to the remote object, it is received by the stub which eventually passes this request to the RRL.
- When the client-side RRL receives the request, it invokes a method called **invoke()**of the object **remoteRef.** It passes the request to the RRL on the server side.
- The RRL on the server side passes the request to the Skeleton (proxy on the server) which finally invokes the required object on the server.
- The result is passed all the way back to the client.

28.3 RMI Registry

- RMI registry is a namespace on which all server objects are placed. Each time the server creates an object, it registers this object with the RMIregistry (using **bind()** or **reBind()**methods). These are registered using a unique name known as **bind name.**
- To invoke a remote object, the client needs a reference of that object. At that time, the client fetches the object from the registry using its bind name (using **lookup()** method).

The following illustration explains the entire process:

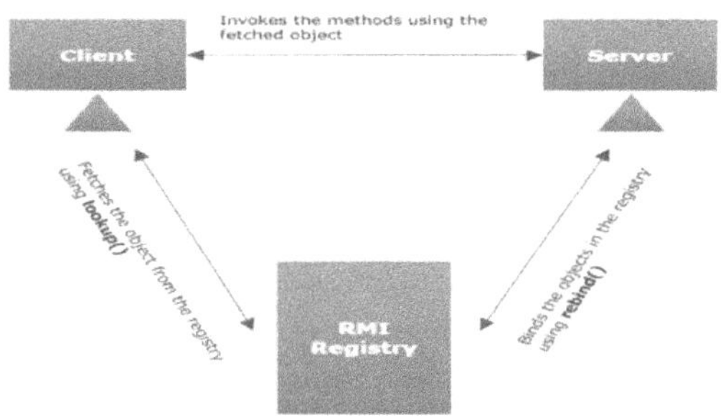

Fig: 28. 2

Goals of RMI

Following are the goals of RMI –

- To minimize the complexity of the application.
- To preserve type safety.
- Distributed garbage collection.
- Minimize the difference between working with local and remote objects.

28.4 Remote Interface

The interfaces and classes that are responsible for specifying the remote behavior of the RMI system are defined in the java.rmi package hierarchy. The following figure shows the relationship between several of these interfaces and classes.

Fig: 28. 3

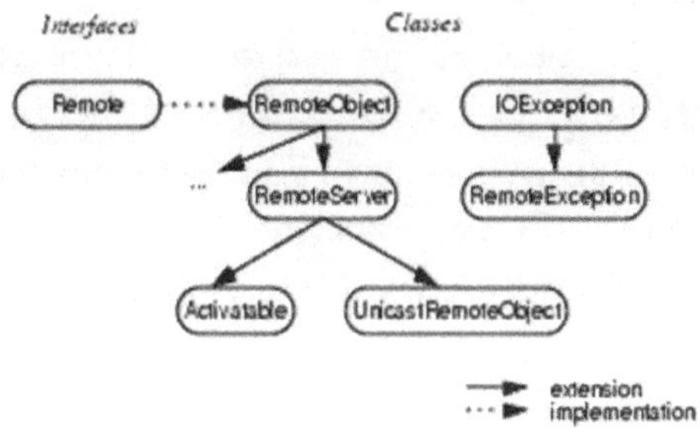

The java.rmi.Remote Interface

In RMI, a remote interface is an interface that declares a set of methods that may be invoked from a remote Java virtual machine. A remote interface must satisfy the following requirements:

- A remote interface must at least extend, either directly or indirectly, the interface **java.rmi.Remote.**
- Each method declaration in a remote interface or its super-interfaces must satisfy the requirements of a remote method declaration as follows:
  1. A remote method declaration must include the exception **java.rmi.RemoteException** (or one of its superclasses such as java.io.IOException or java.lang.Exception) in its throws clause, in addition to any application-specific exceptions (note that application specific exceptions do not have to extend java.rmi.RemoteException).
  2. In a remote method declaration, a remote object declared as a parameter or return value (either declared directly in the parameter list or embedded within a non-remote object in a parameter) must be declared as the remote interface, not the implementation class of that interface.

The interface java.rmi.Remote is a marker interface that defines no methods:

```
public interface Remote {}
```

For example, the following interface BankAccount defines a remote interface for accessing a bank account. It contains remote methods to deposit to the account, to get the account balance, and to withdraw from the account:

```
public interface BankAccount extends java.rmi.Remote {
    public void deposit(float amount)
        throws java.rmi.RemoteException;
    public void withdraw(float amount)
        throws OverdrawnException, java.rmi.RemoteException;
    public float getBalance()
        throws java.rmi.RemoteException;
}
```

28.5 The Naming Class

The Naming class provides methods for storing and obtaining references to remote objects in a remote object registry. Each method of the Naming class takes as one of its arguments a name that is a **java.lang.String** in URL format (without the scheme component) of the form:

```
//host:port/name
```

where host is the host (remote or local) where the registry is located, port is the port number on which the registry accepts calls, and where name is a simple string uninterpreted by the registry. Both host and port are optional. If host is omitted, the host defaults to the local host

Binding a name for a remote object is associating or registering a name for a remote object that can be used at a later time to look up that remote object. A remote object can be associated with a name using the Naming class's bind or rebind methods.

Once a remote object is registered (bound) with the RMI registry on the local host, callers on a remote (or local) host can lookup the remote object by name, obtain its reference, and then invoke remote methods on the object.

28.6 RMI Security Manager Class

```
java.lang.Object
  |
  +----java.lang.SecurityManager
          |
          +----java.rmi.RMISecurityManager
```

- The security manager is an abstract class that allows applications to implement a security policy.
- It allows an application to determine, before performing a possibly unsafe or sensitive operation, what the operation is and whether the operation is being performed by a class created via a class loader rather than installed locally.

- Classes loaded via a class loader (especially if they have been downloaded over a network) may be less trustworthy than classes from files installed locally. The application can allow or disallow the operation.

The **SecurityManager** class contains many methods with names that begin with the word check. These methods are called by various methods in the Java libraries before those methods perform certain potentially sensitive operations. The invocation of such a check method typically looks like this:

```
SecurityManager security = System.getSecurityManager();
   if (security != null) {
      security.checkXXX(argument, . . . );
   }
```

The security manager is thereby given an opportunity to prevent completion of the operation by throwing an exception. A security manager routine simply returns if the operation is permitted, but throws a **SecurityException** if the operation is not permitted. The only exception to this convention is **checkTopLevelWindow,** which returns a boolean value.

The current security manager is set by the **setSecurityManager** method in class System. The current security manager is obtained by the **getSecurityManager** method.

The default implementation of each of the **checkXXX** methods is to assume that the caller does not have permission to perform the requested operation.

28.7 RMI Exceptions

1. Exceptions during Remote Object Export

When a remote object class is created that extends **UnicastRemoteObject,** the object is exported, meaning it can receive calls from external Java virtual machines and can be passed in an RMI call as either a parameter or return value. An object can either be exported on an anonymous port or on a specified port. For objects not extended from UnicastRemoteObject, the **java.rmi.server.UnicastRemoteObject.exportObject** method is used to explicitly export the object.

java.rmi.StubNotFoundException

Name collision with class of same name as stub causes one of these errors:

- Stub can't be instantiated
- Stub not of correct class

2. Exceptions during RMI Call

Table: 28. 1

Exceptions	Context
java.rmi.UnknownHostException	Unknown host.
java.rmi.ConnectException	Connection refused to host.
java.rmi.ConnectIOException	I/O error creating connection.
java.rmi.MarshalException	I/O error marshaling transport header, marshaling call header, or marshaling arguments.
java.rmi.NoSuchObjectException	Attempt to invoke a method on an object that is no longer available.

java.rmi.StubNotFoundException	Remote object not exported.
java.rmi.activation.ActivateFailedException	Thrown by RMI runtime when activation fails during a remote call to an activatable object

3. Exceptions or Errors during Return

java.rmi.UnmarshalException

Corrupted stream leads to either an I/O or protocol error when:

- Marshaling return header
- Checking return type
- Checking return code
- Unmarshaling return

Table: 28. 2

Exceptions	Context
java.rmi.UnexpectedException	An exception not mentioned in the method signature occurred (excluding runtime exceptions). The UnexpectedException exception object contains the underlying exception that was thrown by the server
java.rmi.ServerError	Any error that occurs while the server is executing a remote method. The ServerError exception object contains the underlying error that was thrown by the server.
java.rmi.ServerException	This exception is thrown as a result of a remote method invocation when a RemoteException is thrown while processing the invocation on the server, either while unmarshalling the arguments or executing the remote method itself.
java.rmi.ServerRuntimeException	This exception is not thrown by servers running Java 2 SDK, Standard Edition, v1.2-compatible versions. A RuntimeException is propagated to clients intact.

4. Naming Exceptions

The following table lists the exceptions specified in methods of the **java.rmi.Naming** class and the **java.rmi.registry.Registry** interface.

Table: 28. 3

Exceptions	Context
java.rmi.AccessException	Operation disallowed. The registry restricts bind, rebind, and unbind to the same host. The lookup operation can originate from any host.
java.rmi.AlreadyBoundException	Attempt to bind a name that is already bound.
java.rmi.NotBoundException	Attempt to look up a name that is not bound.
java.rmi.UnknownHostException	Attempt to contact a registry on an unknown host.

5. Activation Exceptions

The following table lists the exceptions that can be thrown in activities involving **activatable** objects. The activation API is in the package **java.rmi.activation.**

Exceptions	Context
java.rmi.activation.ActivateFailedException	Thrown by RMI runtime when activation fails during a remote call to an activatable object
java.rmi.activation.ActivationException	General exception class used by the activation interfaces and classes.
java.rmi.activation.UnknownGroupException	Thrown by methods of the activation classes and interfaces when the ActivationGroupID parameter or ActivationGroupID in an ActivationGroupDesc parameter is invalid.
java.rmi.activation.UnknownObjectException	Thrown by methods of the activation classes and interfaces when the ActivationID parameter is invalid.

Table: 28. 4

28.8 Creating RMI Client and Server classes:

Method is invoked by client on the server using RMI

Fig: 28. 4

RMI is used to communicate between two running java applications on different JVM (Java Virtual Machines). The main motive behind RMI implementation is to invoke one method running on different JVM. The JVM Application, in which an invoked method is running out, is called **RMI Server,** where as the invoking application running on the different JVM is called **RMI Client.**

- **Remote Obj / RMI Server:** It is an object that exposes methods which can be invoked by Remote Client using RMI.
- **RMI Client:** It is an object that invokes remote methods on an RMI Server.
- **Stub:** A stub is proxy that stands for RMI Server on client side and handles remote method invocation on behalf of RMI Client.
- **Skeleton:** It is a proxy that stands for RMI client on server side and handles remote method invocation on RMI Server on behalf of client.
- **Registry Service:** A Registry Service is an application that provides the facility of registration & lookup of Remote stub. A Registry Service provides location transparency of Remote Object to RMI Client.

java.rmi.Remote

Interface and each remote method declared in the remote interface have to throw

```
import java.rmi.*;
public interface HelloInterface          extends Remote {
        public String say(String msg, int a, int b) throws RemoteException;
          }
```

java.rmi.RemoteException.

Lisiting 1: Define Remote Interface, declaring Remote Method

Let's define a class for representing remote object. This class must implement remote interface created in the first point and this class is also responsible for created Stub and Skeleton objects.

In order to provide functionality of creating stub and skeleton object, this class must be extended by **UnicastRemoteObject** and this class must have a default constructor throwing **RemoteException.**

Listing 2: Defining Hello class that extends and implements **UnicastRemoteObject and HelloInterface**

```
import java.rmi.*;
import java.rmi.server.*;
public class Hello
    extends UnicastRemoteObject
     implements HelloInterface {
        private String message;
        public Hello(String msg) throws RemoteException {
                message = msg;
                }
public String say(String m, int a, int b) throws RemoteException {
        String str = "";
        switch(m.charAt(0)){
                case 'a':
                        str = "Addtion : " + add(a,b);
                            break;
                 case 's':
                        str = "Substraction : " + sub(a,b);
                        break;
                case 'm':
                        str = "Multiplication : " + mul(a,b);
                        break;
                   case 'd':
                        str = "Division : " + div(a,b);
                        break;

                default:
                        str = "Wrong Operation";
}
                        return str;
                }
```

Conti.....

```
public int add(int a, int b){
                return a+b;
        }
        public int sub(int a, int b){
                return a-b;
        }
        public int mul(int a, int b){
                return a*b;
        }
        public int div(int a, int b){
                return a/b;
        }
}
```

Listing 3: Define a RMI Server Class

```
import java.rmi.*;
import java.rmi.server.*;
public class HelloServer {
        public static void main(String argv[]) {
                try {
                        Naming.rebind ("Hello", new Hello ("Hello, world!"));
                        System.out.println ("Hello Server is ready.");
                } catch (Exception e) {
                  System.out.println ("Hello Server failed: " + e);
                }
        }
}
```

Listing 4: Define a RMI Client Class

```
import java.rmi.*;
public class HelloClient {
        public static void main(String args[]) {
                try {
if (args.length < 0) {
        System.err.println("usage: java HelloClient string ...\n");
        System.exit(1);
                }
HelloInterface hello = (HelloInterface)Naming.lookup("//localhost/Hello");
                        System.out.println(hello.say(args[0],
Integer.parseInt(args[1]), Integer.parseInt(args[2])));
                }
                catch (Exception e) {
                        System.out.println("Helloclient exception: " + e);
                }
        }   }
```

Listing 5: Define how to register a server in Registry if RMI Registry run on non default port

```
import java.rmi.AlreadyBoundException;
import java.rmi.RemoteException;
import java.rmi.registry.LocateRegistry;
import java.rmi.registry.Registry;

public class RegisterServer {
        public static void main(String[]args){
                try{
                        System.out.println("Createing A RMI Server..");
                        ServerClass serverObject = new ServerClass();
                        System.out.println("Register created RMI Server..");
Registry registerObj = LocateRegistry.getRegistry("localhost", 5000);
//The next statement will throw AlreadyBoundException, that will be caught by
catch statement of class AlreadyBoundException
                        registerObj.bind("serverclass", serverObject);
                        System.out.println("Stub Registered, Server is ready..");
                }
                catch(RemoteException e){
                        e.printStackTrace();
                }
                catch(AlreadyBoundException e){
                        e.printStackTrace();
                }
        }   }
```

Listing 6: If RMI Registry run on non default port, define a RMI Client

```
import java.net.MalformedURLException;
import java.rmi.Naming;
import java.rmi.NotBoundException;
import java.rmi.RemoteException;

public class MyClient {
public static void main(String[]args){
        try {
/*
                * Following statement may throw three kinds of exceptions:
                * @ MalformedURLException
                * @ RemoteException
                * @ NotBoundException
                *
                */
Naming.lookup("rmi://localhost:3000/serverclass");
        } catch (MalformedURLException e) {
                // TODO Auto-generated catch block
                e.printStackTrace();
        } catch (RemoteException e) {
                // TODO Auto-generated catch block
                e.printStackTrace();
        } catch (NotBoundException e) {
                // TODO Auto-generated catch block
                e.printStackTrace();
        }
}
}
```

28.9 RMI

The RMI (Remote Method Invocation) is an API that provides a mechanism to create distributed application in java. The RMI allows an object to invoke methods on an object running in another JVM.

The RMI provides remote communication between the applications using two objects **stub and skeleton.**

Understanding stub and skeleton

RMI uses stub and skeleton object for communication with the remote object.

A remote object is an object whose method can be invoked from another JVM. Let's understand the stub and skeleton objects:

Stub

The stub is an object, acts as a gateway for the client side. All the outgoing requests are routed through it. It resides at the client side and represents the remote object. When the caller invokes method on the stub object, it does the following tasks:

1. It initiates a connection with remote Virtual Machine (JVM),
2. It writes and transmits (marshals) the parameters to the remote Virtual Machine (JVM),
3. It waits for the result
4. It reads (unmarshals) the return value or exception, and
5. It finally, returns the value to the caller.

skeleton

The skeleton is an object, acts as a gateway for the server side object. All the incoming requests are routed through it. When the skeleton receives the incoming request, it does the following tasks:

1. It reads the parameter for the remote method
2. It invokes the method on the actual remote object, and
3. It writes and transmits (marshals) the result to the caller.

In the Java 2 SDK, an stub protocol was introduced that eliminates the need for skeletons.

In the Java 2 SDK, an stub protocol was introduced that eliminates the need for skeletons.

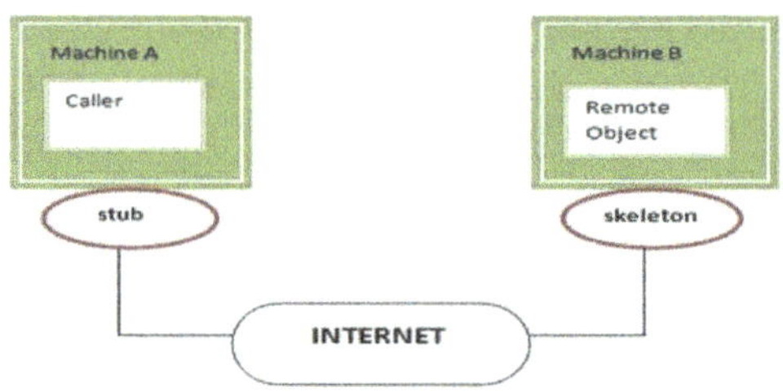

Fig: 28. 5

Understanding requirements for the distributed applications

If any application performs these tasks, it can be distributed application.

1. The application need to locate the remote method
2. It need to provide the communication with the remote objects, and
3. The application need to load the class definitions for the objects.

The RMI applications have all these features, so it is called the distributed application.

Java RMI Example

There is given the 6 steps to write the RMI program.

1. Create the remote interface
2. Provide the implementation of the remote interface
3. Compile the implementation class and create the stub and skeleton objects using the **rmic** tool
4. Start the registry service by **rmiregistry** tool
5. Create and start the remote application
6. Create and start the client application

In this example, we have followed all the 6 steps to create and run the rmi application. The client applications need only two files, remote interface and client application. In the **rmi** application, both client and server interact with the remote interface. The client application invokes methods on the proxy object; RMI sends the request to the remote JVM. The return value is sent back to the proxy object and then to the client application.

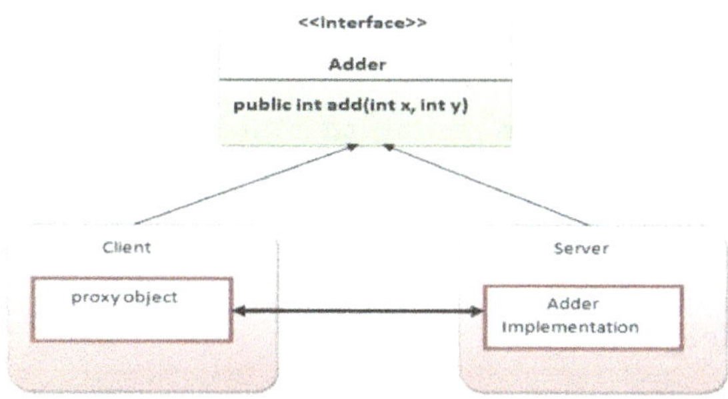

Fig: 28. 6

1) Create the remote interface

For creating the remote interface, extend the Remote interface and declare the **RemoteException** with all the methods of the remote interface. Here, we are creating a remote interface that extends the Remote interface. There is only one method named add() and it declares RemoteException.

```
import java.rmi.*;
public interface Adder extends Remote{
public int add(int x,int y)throws RemoteException;
}
```

2) Provide the implementation of the remote interface

Now provide the implementation of the remote interface. For providing the implementation of the Remote interface, we need to

- Either extend the **UnicastRemoteObject** class,
- or use the **exportObject()** method of the UnicastRemoteObject class

In case, you extend the **UnicastRemoteObject** class, you must define a constructor that declares **RemoteException.**

```
import java.rmi.*;
import java.rmi.server.*;
public class AdderRemote extends UnicastRemoteObject implements Ad
der{
AdderRemote()throws RemoteException{
super();
}
public int add(int x,int y){return x+y;}
}
```

3) create the stub and skeleton objects using the rmic tool.

Next step is to create stub and skeleton objects using the rmi compiler. The **rmic** tool invokes the RMI compiler and creates stub and skeleton objects.

```
rmic AdderRemote
```

4) Start the registry service by the rmiregistry tool

Now start the registry service by using the **rmiregistry** tool. If you don't specify the port number, it uses a default port number. In this example, we are using the port number 5000.

```
rmiregistry 5000
```

5) Create and run the server application

Now **rmi** services need to be hosted in a server process. The Naming class provides methods to get and store the remote object.

```
import java.rmi.*;
import java.rmi.registry.*;
public class MyServer{
public static void main(String args[]){
try{
Adder stub=new AdderRemote();
Naming.rebind("rmi://localhost:5000/sonoo",stub);
}catch(Exception e){System.out.println(e);}
}
}
```

6) Create and run the client application

At the client we are getting the stub object by the lookup() method of the Naming class and invoking the method on this object. In this example, we are running the server and client applications, in the same machine so we are using localhost. If you want to access the remote object from another machine, change the localhost to the host name (or IP address) where the remote object is located.

```
import java.rmi.*;
public class MyClient{
public static void main(String args[]){
try{
Adder stub=(Adder)Naming.lookup("rmi://localhost:5000/sonoo");
System.out.println(stub.add(34,4));
}catch(Exception e){}
}
}
```

28.10 RMI over IOP

RMI-IIOP (read as "RMI over IIOP") denotes the Java Remote Method Invocation (RMI) interface over the Internet Inter-Orb Protocol (IIOP), which delivers Common Object Request Broker Architecture (CORBA) distributed computing capabilities to the Java platform.

It was initially based on two specifications: the Java Language Mapping to OMG IDL, and CORBA/IIOP 2.3.1. With features inherited from CORBA, software components that work together can be written in multiple computer languages and run on multiple computers.

In other words, it supports multiple platforms and can make remote procedure calls to execute, subroutines on another computer as defined by RMI.

CHAPTER-28

JAVABEAN

What is Javabean?

A JavaBean is a specially constructed Java class written in the Java and coded according to the JavaBeans API specifications.

Following are the unique characteristics that distinguish a JavaBean from other Java classes

- It provides a default, no-argument constructor.
- It should be **serializable** and that which can implement the Serializable interface.
- It may have a number of properties which can be read or written.
- It may have a number of **"getter" and "setter"** methods for the properties.

29.1 JavaBeans Properties

A JavaBean property is a named attribute that can be accessed by the user of the object. The attribute can be of any Java data type, including the classes that you define.

A JavaBean property may be **read, write, read only,** or **write only.** JavaBean properties are accessed through two methods in the JavaBean's implementation class

Table: 29. 1

S.No	Method & Description
1	get**PropertyName**() For example, if property name is *firstName*, your method name would be **getFirstName()** to read that property. This method is called accessor.
2	set**PropertyName**() For example, if property name is *firstName*, your method name would be **setFirstName()** to write that property. This method is called mutator.

A read-only attribute will have only a **getPropertyName()** method, and a write-only attribute will have only a **setPropertyName()** method.

JavaBeans Example

```
package com.tutorialspoint;

public class StudentsBean implements java.io.Serializable {
  private String firstName = null;
  private String lastName = null;
  private int age = 0;

  public StudentsBean() {
  }
  public String getFirstName(){
    return firstName;
  }
public String getLastName(){
    return lastName;
  }
  public int getAge(){
    return age;
  }
public void setFirstName(String firstName){

    this.firstName = firstName;

  }

  public void setLastName(String lastName){

    this.lastName = lastName;

  }

  public void setAge(Integer age){

    this.age = age;

  }

}
```

29.2 Accessing JavaBeans

The **useBean** action declares a JavaBean for use in a JSP. Once declared, the bean becomes a scripting variable that can be accessed by both scripting elements and other custom tags used in the JSP. The full syntax for the useBean tag is as follows

```
<jsp:useBean id = "bean's name" scope = "bean's scope" typeSpec/>
```

Here values for the scope attribute can be a **page, request, session** or **application basedon** your requirement. The value of the **id** attribute may be any value as a long as it is a unique name among other **useBean declarations** in the same JSP.

Following example shows how to use the useBean action

```
<html>
  <head>
    <title>useBean Example</title>
  </head>

  <body>
    <jsp:useBean id = "date" class = "java.util.Date" />
    <p>The date/time is <%= date %>
  </body>
</html>
```

29.3 Accessing JavaBeans Properties

Along with **<jsp:useBean...>** action, you can use the **<jsp:getProperty/>** action to access the get methods and the **<jsp:setProperty/>** action to access the set methods. Here is the full syntax –

```
<jsp:useBean id = "id" class = "bean's class" scope = "bean's scope">
  <jsp:setProperty name = "bean's id" property = "property name"
    value = "value"/>
  <jsp:getProperty name = "bean's id" property = "property name"/>
  ...........
</jsp:useBean>
```

The name attribute references the id of a JavaBean previously introduced to the JSP by the useBean action. The property attribute is the name of the **get** or the **set** methods that should be invoked.

Following example shows how to access the data using the above syntax:

```
<html>
  <head>
    <title>get and set properties Example</title>
  </head>

  <body>
    <jsp:useBean id = "students" class = "com.tutorialspoint.StudentsBean">
      <jsp:setProperty name = "students" property = "firstName" value =
"Zara"/>
      <jsp:setProperty name = "students" property = "lastName" value = "Ali"/>
      <jsp:setProperty name = "students" property = "age" value = "10"/>
    </jsp:useBean>
<p>Student First Name:
      <jsp:getProperty name = "students" property = "firstName"/>
    </p>

    <p>Student Last Name:
      <jsp:getProperty name = "students" property = "lastName"/>
    </p>
    <p>Student Age:
      <jsp:getProperty name = "students" property = "age"/>
    </p>

  </body>
```

29.4 Events

An events in java is an object that is created when something changes within a graphical user interface. If a user clickes on a **button, clickes** on a combo box, or types characters into a text field, etc., then an events triggers, creating the relevant event object. This behavior is part of Java's Event handling mechanism and is included in the **Swing GUI library.**

For **example**, let say we have **JButton.**

If a user clicks on the JButton, a button click event is triggered, the event will be created, and it will be sent to the relevant event listener (in this case, the ActionListener). The relevant listener will have implemented code that determines the action to take when the event occurs.

How Events Work?

Event handling in Java is comprised of two key elements:

1. The event source, which is an object that is created when an event occurs. Java provides several types of these event sources, discussed in the section Types of Events below.
2. The event listener, the object that "listens" for events and processes them when they occur.

There are several types of events and listeners in Java: each type of event is tied to a corresponding listener. For this discussion, let's consider a common type of event, an action event represented by the Java class **ActionEvent,** which is triggered when a user clicks a button or the item of a list.

At the user's action, an **ActionEvent** object corresponding to the relevant action is created. This object contains both the event source information and the specific action taken by the user. This event object is then passed to the corresponding **ActionListener** object's method:

```
void actionPerformed(ActionEvent e)
```

This method is executed and returns the appropriate GUI response, which might be to open or close a dialog, download a file, provide a digital signature, or any other of the myriad actions available to users in an interface.

29.5 Types of Events

Here are some of the most common types of events in Java:

- **ActionEvent:** Represents a graphical element is clicked, such as a button or item in a list. Related listener: **ActionListener.**
- **ContainerEvent:** Represents an event that occurs to the GUI's container itself, for example, if a user adds or removes an object from the interface. Related listener: **ContainerListener.**
- **KeyEvent:** Represents an event in which the user presses, types or releases a key. Related listener: **KeyListener.**
- **WindowEvent:** Represents an event relating to a window, for example, when a window is closed, activated or deactivated. Related listener: **WindowListener.**
- **MouseEvent:** Represents any event related to a mouse, such as when a mouse is clicked or pressed. Related listener: **MouseListener.**

Note that multiple listeners and event sources can interact with one another. For example, multiple events can be registered by a single listener, if they are of the same type.

This means that, for a similar set of components that perform the same type of action, one event listener can handle all the events.

A bean class can fire off any type of event, including custom events. As with properties, events are identified by a specific pattern of method names.

```
public void add<Event>Listener(<Event>Listener a)
public void remove<Event>Listener(<Event>Listener a)
```

The listener type must be a descendant of **java.util.EventListener.**

For example, a **Swing JButton** is a bean that fires action events when the user clicks on it. **JButton** includes the following methods (actually inherited from **AbstractButton**), which are the bean pattern for an event.

```
public void addActionListener(ActionListener l);
public void removeActionListener(ActionListener l);
```

29.6 Customization

Customization provides a means for modifying the appearance and behavior of a bean within an application builder so it meets your specific needs. There are several levels of customization available for a bean developer to allow other developers to get maximum benefit from a bean's potential functionality.

The following links are useful for learning about property editors and customizers:

- PropertyEditor interface
- PropertyEditorSupport class
- PropertyEditorManager class
- Customizer interface
- BeanInfo interface

A bean's appearance and behavior can be customized at design time within beans-compliant builder tools. There are two ways to customize a bean:

1. By using a property editor. Each bean property has its own property editor. The NetBeans GUI Builder usually displays a bean's property editors in the Properties window. The property editor that is associated with a particular property type edits that property type.
2. By using customizers. Customizers give you complete GUI control over bean customization. Customizers are used where property editors are not practical or applicable. Unlike a property editor, which is associated with a property, a customizer is associated with a bean.

29.7 Property Editors

A property editor is a tool for customizing a particular property type. Property editors are activated in the Properties window. This window determines a property's type, searches for a relevant property editor, and displays the property's current value in a relevant way.

Property editors must implement the PropertyEditor interface, which provides methods to specify how a property should be displayed in a property sheet. The following figure represents the Properties window containing myBean1 properties:

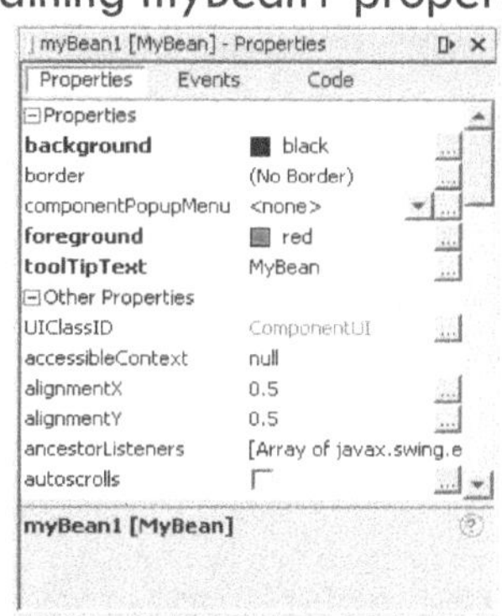

Fig: 29. 1

You begin the process of editing these properties by clicking the property entry. Clicking most of these entries will bring up separate panels. For example, to set up the foreground or background use selection boxes with choices of colors, or press the "..." button to work with a standard **ColorEditor** window. Clicking on the **toolTipText** property opens a **StringEditor** window.

The support class **PropertyEditorSupport** provides a default implementation of the **PropertyEditor** interface. By **subclassing** your property editor from **PropertyEditorSupport**, you can simply override the methods you need.

To display the current property value "sample" within the Properties window, you need to override **isPaintable** to return true. You then must override **paintValue** to paint the current property value in a rectangle in the property sheet. Here's how **ColorEditor** implements **paintValue**:

```
public void paintValue(java.awt.Graphics gfx, java.awt.Rectangle box) {
   Color oldColor = gfx.getColor();
gfx.setColor(Color.black);
gfx.drawRect(box.x, box.y, box.width-3, box.height-3);
gfx.setColor(color);
gfx.fillRect(box.x+1, box.y+1, box.width-4, box.height-4);
gfx.setColor(oldColor);
}
```

To support the custom property editor, override two more methods. Override **supportsCustomEditor** to return true, and then override **getCustomEditor** to return a custom editor instance.

All customizers must:

- Extend **java.awt.Component** or one of its subclasses.
- Implement the **java.beans.Customizer** interface This means implementing methods to register **PropertyChangeListener** objects, and firing property change events at those listeners when a change to the target bean has occurred.
- Implement a default constructor.
- Associate the customizer with its target class via **BeanInfo.getBeanDescriptor.**

29.8 Introspection

Introspection is the automatic process of analyzing a bean's design patterns to reveal the bean's properties, events, and methods. This process controls the publishing and discovery of bean operations and properties. This lesson explains the purpose of introspection, introduces the Introspection API, and gives an example of introspection code.

Purpose of Introspection

A growing number of Java object repository sites exist on the Internet in answer to the demand for centralized deployment of applets, classes, and source code in general. Any developer who has spent time hunting through these sites for licensable Java code to incorporate into a program has undoubtedly struggled with issues of how to quickly and cleanly integrate code from one particular source into an application.

The way in which introspection is implemented provides great advantages, including:

- **Portability -** Everything is done in the Java platform, so you can write components once, reuse them everywhere. There are no extra specification files that need to be maintained independently from your component code. There are no platform-specific issues to contend with. Your component is not tied to one component model or one proprietary platform. You get all the advantages of the evolving Java APIs, while maintaining the portability of your components.
- **Reuse -** By following the JavaBeans design conventions, implementing the appropriate interfaces, and extending the appropriate classes, you provide your component with reuse potential that possibly exceeds your expectations.

29.9 Introspection API

The JavaBeans API architecture supplies a set of classes and interfaces to provide introspection.

The **Introspector** class provides descriptor classes with information about properties, events, and methods of a bean. Methods of this class locate any descriptor information that has been explicitly supplied by the developer through **BeanInfo** classes. Then the Introspector class applies the naming conventions to determine what properties the bean has, the events to which it can listen, and those which it can send.

The following figure represents a hierarchy of the **FeatureDescriptor** classes:

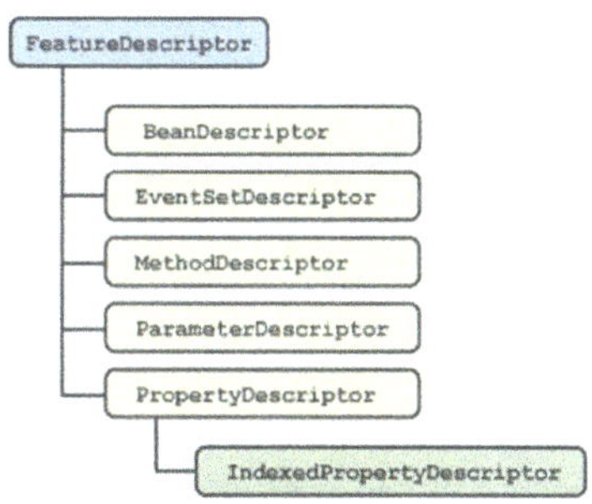

Fig: 29. 2

Introspection Sample Example:

```
import java.beans.BeanInfo;
import java.beans.Introspector;
import
java.beans.IntrospectionException;
import java.beans.PropertyDescriptor;

public class SimpleBean
{
private final String name = "SimpleBean";
private int size;

public String getName()
   {
return this.name;
   }
public int getSize()
   {
return this.size;
```

Conti......

```
public void setSize( int size )
  {
    this.size = size;
  }

public static void main( String[] args )
throws IntrospectionException
  {
    BeanInfo info = Introspector.getBeanInfo( SimpleBean.class
);
for ( PropertyDescriptor pd : info.getPropertyDescriptors() )
System.out.println( pd.getName() );
  }
```

29.10 Bean Persistence

A bean has the property of persistence when its properties, fields, and state information are saved to and retrieved from storage. Component models provide a mechanism for persistence that enables the state of components to be stored in a non-volatile place for later retrieval.

The mechanism that makes persistence possible is called ***serialization.*** Object serialization means converting an object into a data stream and writing it to storage. Any applet, application, or tool that uses that bean can then "reconstitute" it by deserialization. The object is then restored to its original state.

For example, a Java application can serialize a Frame window on a Microsoft Windows machine, the serialized file can be sent with e-mail to a Solaris machine, and then a Java application can restore the Frame window to the exact state which existed on the Microsoft Windows machine.

All beans must persist. To persist, your beans must support serialization by implementing either the **java.io.Serializable** (in the API reference documentation) interface, or the **java.io.Externalizable** (in the API reference documentation) interface. These interfaces offer you the choices of automatic serialization and customized serialization. If any class in a class's inheritance hierarchy implements Serializable or Externalizable, then that class is serializable.

CHAPTER-29

EJB

What is EJB?

EJB is an acronym for enterprise java bean. It is a specification provided by Sun Microsystems to develop secured, robust and scalable distributed applications.

To run EJB application, you need an application server (EJB Container) such as Jboss, Glassfish, Weblogic, Websphere etc. It performs:

- life cycle management,
- security,
- transaction management, and
- Object pooling.

EJB application is deployed on the server, so it is called server side component also. EJB is like COM (Component Object Model) provided by Microsoft. But, it is different from Java Bean, RMI and Web Services.

When use Enterprise Java Bean?

1. Application needs Remote Access. In other words, it is distributed.
2. Application needs to be scalable. EJB applications supports load balancing, clustering and fail-over.
3. Application needs encapsulated business logic. EJB application is separated from presentation and persistent layer.

30.1 Types of Enterprise Java Bean

There are 3 types of enterprise bean in java.

1. **Session Bean**- Session bean contains business logic that can be invoked by local, remote or web service client.
2. **Message Driven Bean**- like Session Bean, it contains the business logic but it is invoked by passing message.
3. **Entity Bean**- It encapsulates the state that can be persisted in the database. It is deprecated. Now, it is replaced with JPA (Java Persistent API).

30.2 Difference between RMI and EJB

Both RMI and EJB, provides services to access an object running in another JVM (known as remote object) from another JVM. The differences between RMI and EJB are given below:

Table: 30.1

RMI	EJB
In RMI, middleware services such as security, transaction management, object pooling etc. need	In EJB, middleware services are provided by EJB Container automatically.

to be done by the java programmer.	
RMI is not a server-side component. It is not required to be deployed on the server.	EJB is a server-side component; it is required to be deployed on the server.
RMI is built on the top of socket programming.	EJB technology is built on the top of RMI.

30.3 EJB and Web service

- In EJB, bean component and bean client both must be written in java language.
- If bean client need to be written in other language such as **.net, php** etc, we need to go with web services (SOAP or REST).

Disadvantages of EJB

- Requires application server
- Requires only java client. For other language client, you need to go for web service.
- Complex to understand and develop EJB applications.

30.4 EJB Container

An EJB container will provide the runtime environment for EJB applications within the application server.

- It is responsible for all the operations of the EJB applications.
- The container acts as an intermediary action between the business logic of the bean and the rest of the world of the enterprise application.
- One or more EJB modules can be installed within a single EJB container.
- The role of EJB container is to perform transactional actions such as,

1. Starting a transaction.
2. Rollback a transaction or commit a transaction.
3. Managing various connection pools for the database resources.
4. Bean's instance variables with corresponding data items which are stored in a database will be synchronized.

Explain the difference between servers and containers.

- The server receives the requests from client and sends the responses to the client.
- If it is a web server or EJB server, the request will be sent to the web container or to the EJB container.
- A web server responds requests over the web.
- The container creates the required objects pertaining to the requests and responses, and invokes the appropriate service methods.
- Containers also provide the communication support, life cycle management, multithreading support and declarative security.
- A container runs inside the web or application server which could produce dynamic content.

Classes

In the real world, you'll often find many individual objects all of the same kind. There may be thousands of other bicycles in existence, all of the same make and model. Each bicycle was built from the same set of blueprints and therefore contains the same components. In object-oriented terms, we say that your bicycle is an instance of the class of objects known as bicycles. A class is the blueprint from which individual objects are created. The following **Bicycle class** is one possible implementation of a bicycle:

```
class Bicycle {
int cadence = 0;
int speed = 0;
int gear = 1;
void changeCadence(int newValue) {
cadence = newValue;
   }
void changeGear(int newValue) {
gear = newValue;
   }
void speedUp(int increment) {
speed = speed + increment;
   }
void applyBrakes(int decrement) {
speed = speed - decrement;
   }
void printStates() {
System.out.println("cadence:" +
cadence + " speed:" +
speed + " gear:" + gear);
   }  }
```

The syntax of the Java programming language will look new to you, but the design of this class is based on the previous discussion of bicycle objects. The fields cadence, speed, and gearrepresent the object's state, and the methods (changeCadence, changeGear, speedUp etc.) define its interaction with the outside world.

You may have noticed that the Bicycle class does not contain a main method. That's because it's not a complete application; it's just the blueprint for bicycles that might be used in an application. The responsibility of creating and using new Bicycle objects belongs to some other class in your application.

Here's a BicycleDemo class that creates two separate Bicycle objects and invokes their methods:

```
class BicycleDemo {
public static void main(String[] args) {

        // Create two different
        // Bicycle objects
        Bicycle bike1 = new Bicycle();
        Bicycle bike2 = new Bicycle();

        // Invoke methods on
        // those objects
bike1.changeCadence(50);
bike1.speedUp(10);
bike1.changeGear(2);
bike1.printStates();

bike2.changeCadence(50);
bike2.speedUp(10);
bike2.changeGear(2);
bike2.changeCadence(40);
bike2.speedUp(10);
bike2.changeGear(3);
bike2.printStates();
    }
}
```

30.5 EJB Interfaces

Interface in java means a group of related methods with empty bodies. EJB have generally 4 interfaces. These are as follows

1) Remote interface:- Remote interface are the interface that has the methods that relate ta particular bean instance. In the Remote interface we have all get methods as given below in the program. This is the interface where the entire business method **go.javax.ejb.Remote** package is used for creating Remote interface.

```
package ejb;

import javax.ejb.Remote;

@Remote
public interface Bean30Remote {

String getMessage();

String getAddress();

String getCompanyname();
}
```

2) Local Interface:-Local interface are the type of interface that are used for making local connections to **EJB.@Local** annotation is used for declaring interface as Local. **javax.ejb.Local** package is used for creating Local interface.

```
package ejb;

import javax.ejb.Local;

@Local
public interface NewSessionLocal {

}
```

3) Home Interface:-Home interface is the interface that has methods that relate to all EJB of a certain class as a whole. The methods which are defined in this Interface are **create ()** methods and find () methods . The **create ()** method allows us to create beans. **find ()** method is used in Entity beans.

```
import javax.ejb.EJBHome;
import javax.ejb.CreateException;
import java.rmi.RemoteException;

public interface Bean30RemoteHome extends EJBHome{
public Bean30Remote create() throws CreateException,
RemoteException;
}
```

4) Localhome Interface:-The local interfaces extend the following interfaces. These interfaces are generally for use by clients

- **javax.ejb.EJBLocalObject -** for the Object interface
- **javax.ejb.EJBLocalHome -** for the Home interface

```
package ejb;

import javax.ejb.EJBLocalHome;
import javax.ejb.CreateException;

public interface NewSessionLocalHome extends EJBLocalHome {
public NewSessionLocal create() throws CreateException;
}
```

30.6 EJB deployment description

Deployment descriptor is the file which tells the EJB server that which classes make up the bean implementation, the home interface and the remote interface. It also indicates the behavior of one EJB with other. The deployment descriptor is generally called as ejb-jar.xml and is in the directory META-INF of the client application. In the example given below our application consists of single EJB.

<ejb-name>secure</ejb-name>:-This is the node that assigns the name to the EJB.

<description>Accessing Database Application</description>:-This node gives the brief description about the Ejb module created.

<session-type>Stateless</session-type>:-This node assigns the Session bean as stateless or stateful. Here stateless means to say accessing Remote interface.

```
<?xml version="1.0" encoding="UTF-8"?>
<application-client version="5"
xmlns="http://java.sun.com/xml/ns/javaee"  xmlns:xsi="http://www.w3.org/2001/X
MLSchema-instance" xsi:schemaLocation="http://java.sun.com/xml/ns/javaee
http://java.sun.com/xml/ns/javaee/application-client_5.xsd">
<description>Accessing Database Application</description>
<display-name>Secure-app-client</display-name><enterprise-beans>
<session>
<ejb-name>secure</ejb-name>
<home>org.glassfish.docs.secure.secureHome</home>
<remote>org.glassfish.docs.secure.secure</remote>
<ejb-class>org.glassfish.docs.secure.secureBean</ejb-class>
<session-type>Stateless</session-type>
</session>
</enterprise-beans>
</application-client>
```

30.7 Session Beans

A session bean is the enterprise bean that directly interacts with the user and contains the business logic of the enterprise application. A session bean represents a single client accessing the enterprise application deployed on the server by invoking its method. An application may contain multiple sessions depending upon the number of users accessing to the application. A session bean makes an interactive session only for a single client and shields that client from complexities just by executing the business task on server side.

For example, whenever a client wants to perform any of these actions such as making a reservation or validating a credit card, a session bean should be used. The session bean decides what data is to be modified. Typically, the session bean uses an entity bean to access or modify data. They implement business logic, business rules, algorithms, and work flows. Session beans are relatively short-lived components. The EJB container may destroy a session bean if its client times out.

A session bean can neither be shared nor can persist (means its value cannot be saved to the database) its value. A session bean can have only one client. As long as the client terminates, session been associated with this client is also terminated and the data associated with this bean is also destroyed.

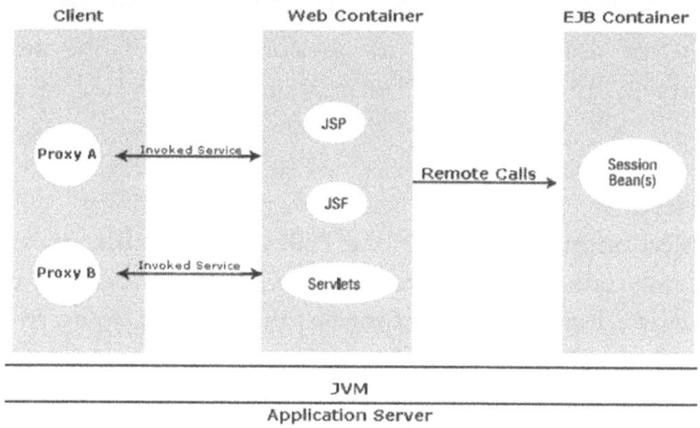

Fig: 30.1

The above figure shows how Session Bean interacts with the clients as well as with the Entity Beans.

Session beans are divided into two parts.

1) Stateless Session Beans:

A stateless session bean does not maintain a conversational state with the client. When a client invokes the methods of a stateless bean, the instance of bean variables may contain a state specific to that client only for the duration of a method invocation. Once the method is finished, the client-specific state should not be retained i.e. the EJB container destroys a stateless session bean.

These types of session beans do not use the class variables (instance variables). So they do not persist data across method invocation and therefore there is no need to passivate the bean's instance. Because stateless session beans can support multiple clients, they provide the better scalability for applications that require large numbers of clients.

2) Stateful Session Beans:

These types of beans use the instance variables that allow the data persistent across method invocation because the instance variables allow persistence of data across method invocation. The client sets the data to these variables which he wants to persist.

A stateful session bean retains its state across multiple method invocations made by the same client. If the stateful session bean's state is changed during a method invocation, then that state will be available to the same client on the following invocation. The state of a client bean is retained for the duration of the client-bean session.

Once the client removes the bean or terminates, the session ends and the state disappears. Because the client interacts with its bean, this state is often called the conversational state.

For example, consider a customer using a debit card at an ATM machine. The ATM could perform various operations like checking an account balance, transferring funds, or making a withdrawal. These operations could be performed one by one, by the same customer. So the bean needs to keep track its state for each of these operations to the same client.

Thus Stateful session beans have the extra overhead for the server to maintain the state than the stateless session bean.

The user interface calls methods of session beans if the user wants to use the functionality of the session bean. Session beans can call to other session beans and entity beans.

When to use session beans:

Generally session beans are used in the following circumstances:

1. When there is only one client is accessing the beans instance at a given time.
2. When the bean is not persistent that means the bean is going to exist no longer.
3. The bean is implementing the web services.

Stateful session beans are useful in the following circumstances:

- What the bean wants to holds information about the client across method invocation.
- When the bean works as the mediator between the client and the other component of the application.
- When the bean have to manage the work flow of several other enterprise beans.

Stateless session beans are appropriate in the circumstances illustrated below:

- If the bean does not contain the data for a specific client.
- If there is only one method invocation among all the clients to perform the generic task.

30.8 Life Cycle of a Stateless Session Bean:

Since the Stateless session bean does not passivate across method calls therefore a stateless session bean includes only two stages. Whether it does not exist or ready for method invocation. A stateless session bean starts its life cycle when the client first obtains the reference of the session bean.

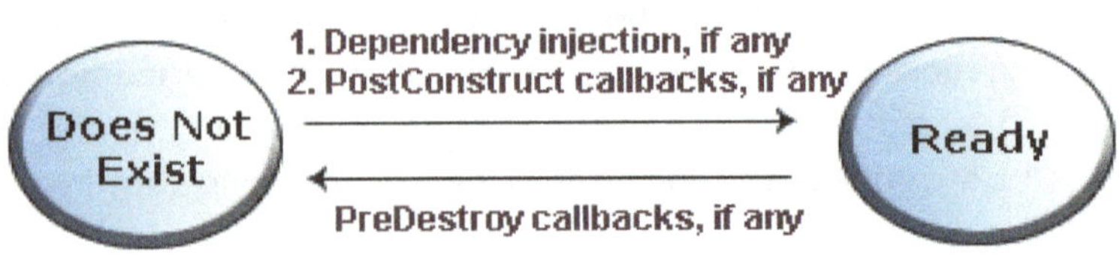

Fig: 30.2

The above figure demonstrates how the Stateless Session Beans are created and destroyed.

30.9 Life Cycle of a Stateful Session Bean:

A Stateful session bean starts its life cycle when the client first gets the reference of a stateful session bean. The container may deactivate a bean while in ready state (Generally the container uses the least recently use algorithm to passivates a bean). In the passivate mechanism the bean moves from memory to secondary memory.

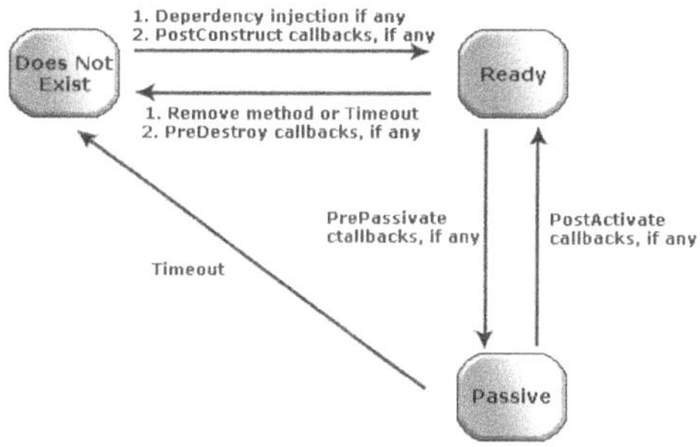

Fig: 30.3

30.10 Entity Java Bean

Entity beans are objects that represent a persistence storage mechanism. The objects can be a customer, product, account etc. Each entity bean has underlying table in a relational database and each row in the table represents the instance of the bean.

Need of entity beans:

1. When it is needed by multiple clients.
2. When any action from the database side.
3. When a transaction is to be performed from the client.
4. When the enforcement of accuracy and integrity is needed.

The types of Entity beans and their uses.

There are two types of entity beans:

1. Container Managed Persistence (CMP)
2. Bean Managed Persistence (BMP)

CMP can be used when:

1. No code for database code is needed (by placing the specific accessibility is provided by XML file).
2. The code reusability is needed.
3. The speed of development by creation of the database access code by the developers.

BMP can be used when:

- Synchronizing the state of the database directly by the container.

- To provide the flexibility for the bean developer to perform the persistence operations.
- The need of the query exceeds the capabilities of EJB QL.
- The persistent storage is a non database system or is a legacy database system that does not supports for CMP.

The steps involved in developing Entity beans.

1. Select which bean is to be used – BMP or CMP entity bean
2. Create the required entity bean.
3. Author the source code for the entity bean, which involves developing the following:
   - The component interface(s)
   - The home interface(s)
   - The bean class
   - Any helper classes required by the entity bean
4. Develop the deployment descriptors for the beans.
5. Create the database tables. (If your database tables already exist, you can omit this step)
6. Establish the Object Relation mapping.
7. Compile and archive all the EJB components into a JAR file.
8. Archive the complete application into an EAR file.
9. Make the deployment of the application onto the J2EE engine.
10. Start the application.

30.11 The Life Cycle of an Entity Bean

To understand how to best develop entity beans, it is important to understand how the container manages them. The EJB specification defines just about every major event in an entity bean's life, from the time it is instantiated to the time it is garbage collected. This is called the life cycle, and it provides the bean developer and EJB vendors with all the information they need to develop beans and EJB servers that adhere to a consistent protocol.

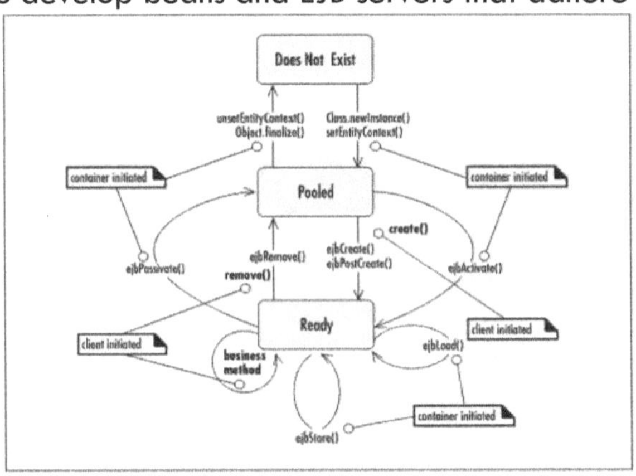

Fig: 30.4

Entity bean life cycle

We will examine the life cycle of an entity bean and identify the points at which the container would call each of the methods described in the EntityBean interface. Bean instances must implement the EntityBean interface, which means that invocations of the callback methods are invocations on the bean instance itself.

1. **Does Not Exist**

   The bean instance begins life as a collection of files. Included in that collection are the bean's deployment descriptor, the remote interface, the home interface, the primary key, and all the supporting classes generated at deployment time. At this stage, no instance of the bean exists.

2. **The Pooled State**

   When the EJB server is started, it reads the bean's files and instantiates several instances of the bean, which it places in a pool. The instances are created by calling the **Class.newInstance**() method on the bean class. The **newInstance()** method creates an instance using the default constructor, which has no arguments.

3. **The Ready State**

   When a bean instance is in the Ready State, it can accept client requests. A bean instance moves to the Ready State when the container assigns it to an EJB object. This occurs under two circumstances: when a new entity bean is being created or when the container is activating an entity.

i) Transitioning from the Pooled state to the Ready State via creation

When a client application invokes the create() method on an EJB home, several operations must take place before the EJB container can return a remote reference (EJB object) to the client. First, an EJB object must be created on the EJB server. Once the EJB object is created, a bean instance is taken from the instance pool and assigned to the EJB object. Next, the create() method, invoked by the client, is delegated to its corresponding **ejbCreate**() method on the bean instance. After the **ejbCreate**() method completes, **aprimary** key is created. In container-managed persistence, the container instantiates and populates the key automatically; in bean-managed persistence, the bean instantiates and populates the primary key and returns it from the **ejbCreate**() method. The key is embedded in the EJB object, providing it with identity. Once the EJB object has identity, the bean instance's **EntityContext** can access information specific to that EJB object, including the primary key and its own remote reference. While the **ejbCreate**() method is executing, the properties, security, and transactional information is available.

ii) Transitioning from the Pooled state to the Ready State via a find method

When a find method is executed, each bean that is found will be realized by transitioning an instance from the Pooled state to the Ready State. When a bean is found, it is assigned to an EJB object and its remote reference is returned to the client. A found bean follows the same protocol as a passivity bean; it is activated when the client invokes a business method. A found bean can be considered to be a passivity bean and will move into the Ready State through activation as described in the next section.

iii) Transitioning from the Pooled state to the Ready State via activation

When a bean instance is activated, it leaves the instance pool (the Pooled State) to be assigned to an EJB object. When a bean instance makes this transition, the bean instance is first assigned to an EJB object. Once assigned to the EJB object, the **ejbActivate()** method is called--the instance's EntityContext can now provide information specific to the EJB object, but it cannot provide security or transactional information. This callback method can be used in the bean instance to reobtain any resources or perform some other work needed before servicing the client.

When an entity bean instance is activated, nonpersistent fields contain arbitrary values (dirty values) and must be reinitialized in the ejbActivate() method.

In container-managed persistence, persistent fields (container-managed fields) are automatically synchronized with the database after ejbActivate() is invoked and before a business method can be serviced by the bean instance. The order in which these things happen is as follows:

1. **ejbActivate()** is invoked on the bean instance.
2. Persistent fields are synchronized automatically.
3. **ejbLoad()** notifies the bean that its persistent fields have been synchronized.
4. Business methods are invoked as needed.
5. In bean-managed persistence, persistent fields are synchronized by the **ejbLoad()** method after **ejbActivate()** has been called and before a business method can be invoked. Here is the order of operations in bean-managed persistence:

- **ejbActivate()** is invoked on the bean instance.
- **ejbLoad()** is called to let the bean synchronize its persistent fields.
- Business methods are invoked as needed.

IV) Transitioning from the Ready State to the Pooled state via passivation

The most fundamental thing to remember is that, for entity beans, passivation is simply a notification that the instance is about to be disassociated from the EJB object. Unlike stateful session beans, an entity bean instance's fields are not serialized and held with the EJB object when the bean is passivated. Whatever values are held in the instance's fields when it was assigned to the EJB object will be carried with it to its next assignment.

V) Transitioning from the Ready State to the Pooled state via removal

A bean instance also moves from the Ready State to the Pooled state when it is removed. This occurs when the client application invokes one of the remove() methods on the bean's EJB object or EJB home. With entity beans, invoking a remove method means that the entity's data is deleted from the database.

4. Life in the Ready State

A bean is in the Ready State when it is associated with an EJB object and is ready to service requests from the client. When the client invokes a business method, like **Ship.getName()**, on the bean's remote reference (EJB object stub), the method invocation is received by the EJB object server and delegated to the bean instance. The instance performs the method and returns the results. As long as the bean instance is in the Ready State, it can service all the business methods invoked by the client. Business methods can be called zero or more times in any order.

The **ejbLoad()** and **ejbStore()** methods, which synchronize the bean instance's state with the database, can be called only when the bean is in the Ready State. These methods can be called in any order, depending on the vendor's implementation. Some vendors call **ejbLoad()** before every method invocation and **ejbStore()** after every method invocation, depending on the transactional context. Other vendors call these methods less frequently.

5. End of the Life Cycle

A bean instance's life ends when the container decides to remove it from the pool and allow it to be garbage collected. This happens under a few different circumstances. If the container decides to reduce the number of instances in the pool--usually to conserve resources--it releases one or more bean instances and allows them to be garbage collected. The ability to adjust the size of the instance pool allows the EJB server to manage its resources--the number of threads, available memory, etc.--so that it can achieve the highest possible performance. This behavior is typical of a CTM.

When an EJB server is shut down, most containers will release all the bean instances so that they can be safely garbage collected. Finally, some containers may decide to release an instance that is behaving unfavorably or an instance that has suffered from some kind of unrecoverable error that makes it unstable. A bean could be behaving unfavorably if, for example, it has a bug that causes it to enter an endless loop or create an unusually large number of objects.

30.12 Jar File

JAR stands for **Java ARchive.** It's a file format based on the popular ZIP file format and is used for aggregating many files into one. Although JAR can be used as a general archiving tool, the primary motivation for its development was so that Java applets and their requisite components (.class files, images and sounds) can be downloaded to a browser in a single HTTP transaction, rather than opening a new connection for each piece. This greatly improves the speed with which an applet can be loaded onto a web page and begin functioning.

The JAR format also supports compression, which reduces the size of the file and improves download time still further. Additionally, individual entries in a JAR file may be digitally signed by the applet author to authenticate their origin.

JAR is:

- the only archive format that is cross-platform
- the only format that handles audio and image files as well as class files
- backward-compatible with existing applet code
- an open standard, fully extendable, and written in java
- the preferred way to bundle the pieces of a java applet

30.13 The APPLET tag

Changing the APPLET tag in your HTML page to accomodate a JAR file is simple. The JAR file on the server is identified by the **ARCHIVE** parameter, where the directory location of the jar file should be relative to the location of the html page:

```
<applet code=Animator.class
archive="jars/animator.jar"
width=460 height=160>
<param name=foo value="bar">
</applet>
```

Note that the familiar CODE=myApplet.class parameter must still be present. The CODE parameter, as always, identifies the name of the applet where execution begins. However, the class file for the applet and all of its helper classes are loaded from the JAR file.

The ARCHIVE attribute describes one or more JAR files containing classes and other resources that will be "preloaded". The classes are loaded using an instance of an AppletClassLoader with the given CODEBASE. It takes the **formarchive = archiveList.** The archives in archiveList are separated by ",".

Once the archive file is identified, it is downloaded and separated into its components. During the execution of the applet, when a new class, image or audio clip is requested by the applet, it is searched for first in the archives associated with the applet.

The archive tag may specify multiple JAR files. Each JAR file must be separated by "," (comma). Each file is downloaded in turn:

```
<applet code=Animator.class
archive="classes.jar ,  images.jar ,  sounds.jar"
width=460 height=160>
<param name=foo value="bar">
</applet>
```

There can be any amount of white space between entries within the archive parameter. In addition, the archive tag itself is case-insensitive; it can be lower-case, upper-case, or any combination of lower- and upper-case letters, such as ArCHiVe.

30.14 Executable JAR Files

On Microsoft Windows systems, the Java 2 Runtime Environment's installation program will register a default association for JAR files so that double-clicking a JAR file on the desktop will automatically run it with javaw -jar. Dependent extensions bundled with the application will also be loaded automatically. This feature makes the end-user runtime environment easier to use on Microsoft Windows systems.

The Solaris 2.6 kernel has already been extended to recognize the special "magic" number that identifies a JAR file, and to invoke java -jar on such a JAR file as if it were a native Solaris executable. A application packaged in a JAR file can thus be executed directly from the command line or by clicking an icon on the CDE desktop.

Let us see how to create a .jar file and related commands which help us to work with **.jar files**

1. Create a JAR file: To create a **.jar file** , we can use jar cf command in the following way:

```
jar cf jarfilename inputfiles
```

Here, cf represents create the file. For example , assuming our package pack is available in **C:\directory** , to convert it into a jar file into the pack.jar , we can give the command as:

```
C:\> jar cf pack.jar pack
```

2. Viewing a JAR file: To view the contents of .jar files, we can use the command as:

```
jar tf jarfilename
```

Here , **tf** represents table view of file contents. For example, to view the contents of our pack.jar file, we can give the command:

```
C:/> jar tf pack.jar
```

Now, the contents of pack.jar are displayed as:

where class1 , class2 etc are the classes in the package pack. The first two entries represent that there is a manifest file created and added to pack.jar. The third entry represents the sub-directory with the name pack and the last two represent the files name in the directory pack.

```
META-INF/
META-INF/MANIFEST.MF
pack/
pack/class1.class
pack/class2.class
..
..
```

3. Extracting a JAR file: To extract the files from a .jar file, we can use:

```
jar xf jarfilename
```

Here, xf represents extract files from the jar files. For example, to extract the contents of our pack.jar file, we can write:

```
C:\> jar xf pack.jar
```

This will create the following directories in C:\

```
META-INF

pack // in this directory , we can see class1.class and class2.class.
```

4. Updating a JAR File The Jar tool provides a 'u' option which you can use to update the contents of an existing JAR file by modifying its manifest or by adding files. The basic command for adding files has this format:

```
jar uf jar-file input-file(s)
```

here uf represent update jar file. For example , to update the contents of our pack.jar file, we can write:

```
C:\>jar uf pack.jar
```

5. Running a JAR file: In order to run an application packaged as a JAR file (requires the Main-class manifest header) , following command can be used:

```
C:\>java -jar pack.jar
```